Cities and Nature

Cities and Nature fosters high-quality multi-disciplinary research addressing the interface between cities and the natural environment. It provides a valuable source of relevant knowledge for researchers, planners and policy-makers. The series welcomes empirically based, cutting-edge and theoretical research in urban geography, urban planning, environmental planning, urban ecology, regional science and economics. It publishes peer-reviewed edited and authored volumes on topics dealing with the urban and the environment nexus, including: spatial dynamics of urban built areas, urban and peri-urban agriculture, urban greening and green infrastructure, environmental planning, urban forests, urban ecology, regional dynamics and landscape fragmentation.

Indexed in Scopus!

Joseph Adeniran Adedeji

Ecological Urbanism of Yoruba Cities in Nigeria

An Ecosystem Services Approach

Foreword by Roman J. M. Lenz

Springer

Joseph Adeniran Adedeji
Department of Architecture
Federal University of Technology
Akure, Nigeria

ISSN 2520-8306 ISSN 2520-8314 (electronic)
Cities and Nature
ISBN 978-3-031-34687-3 ISBN 978-3-031-34688-0 (eBook)
https://doi.org/10.1007/978-3-031-34688-0

This Springer imprint is published by the registered company Springer Nature Switzerland AG
The registered company address is: Gewerbestrasse 11, 6330 Cham, Switzerland

For the glory of God Almighty

For Temileye, Adeyemo, Adeyemi, Adefunke, and Adewumi

Adedeji

Foreword

The subject of this book deals with increasingly important issues, as green infrastructure and their ecosystem services are basic contributions for the wellbeing of inhabitants of landscapes, especially of cities. In addition, climate change as well as biodiversity losses are challenging architectural, ecological, and social contributions to mitigate negative effects. These themes are central to the discourse of ecological urbanism with particular focus on how humans use urban green spaces to explore sustainable connection with the non-human nature in cities. The book uses this perspective of ecological urbanism to study Yoruba cities in Nigeria to unpack how Indigenous Knowledge System (IKS) manifests in everyday lifestyles to shape urban spaces. Through the concept of ecosystem services, it argues that symbiotic relationships that are culturally enhanced exist between humans and non-humans in cities to make them truly ecological.

The Yoruba people have a pre-colonial tradition of closely knit urban settlements where green spaces are conserved as spiritualized groves and secular parks and gardens. This book tells the story of the cosmological worldviews of the Yoruba people in relation to nature, green spaces, and myths, and how such religio-cultural practices could serve as the bases for suggesting building and planning guidelines and frame ecological qualities of cities. The results of the study carried out by Joseph and narrated in this book give credence to the notion that IKS is potent, enduring, and has capacity to inform policy guidelines. Exploring the connections of green spaces, ecosystem services, nature, and cultural ethos with the idea of IKS could extend the frontiers of knowledge on the multispecies character of cities, as Joseph has carried out with focus on the Yoruba people. Throughout the book, Joseph constantly supports the claims of the IKS of the Yoruba people on ecosystem services of nature elements in the cities with scientific literature. The strength of the book lies in its multidisciplinary approach that combines urban ecology, anthropology, ethnobiology, urban history, and socioeconomic perspectives to formulate frameworks that could guide landscape design and urban planning policies for ecological urbanism. Joseph contextualized ecological urbanism of Yoruba cities as consisting of pro-environmental practices, culture, "local wisdom, practices, ideologies, customs, tradition, norms, habits" as a whole range of indigenous wisdom of Yoruba people as form-giving factors of Yoruba cities that promote human ecology.

Joseph Adedeji has worthily took on the responsibility of writing this monograph in view of his professional background as an architect and urban designer who pursues the goal of sustainable urbanism by focusing on landscape hermeneutics and how to practically deploy such knowledge as a feedback strategy into design process. This transverse of theory to practice and vice versa is not only appropriate but highly indispensable for making cities ecological, including co-evolution of urban spaces and pro-ecological behaviours. The ethnographic method that Joseph engaged in his study of ecological urbanism to produce this monograph makes the narratives to be evidently real, being located in everyday life of the participants in the study. The formulated ecological planning model, framework, and design algorithms are situated within the context of sustainable urbanism.

The book would be a necessary companion for researchers in landscape architecture and the mentioned fields of knowledge for further conversations and as a base for policy formulation in governing urban spaces in Yoruba land with areas of possible applications in the Global South, where similarities are confirmed to exist in how humans relate with other forms of life in cities.

Faculty Environment Design Therapy,
University of Applied Sciences
(Hochschule für Wirtschaft und Umwelt, HfWU)
Nuertingen-Geislingen, Germany
January 2023

Roman J. M. Lenz

Preface

This book is an ethnographic text on the phenomenology of ecosystem services (ES) of green infrastructure (GI) and nature in cities. It is motivated by the increasing need to awaken the human consciousness about experiences of cityscapes and urban life in this age of the Anthropocene. The goals of writing the book are to present eco-cultural strategies for urban landscape design, stimulate reconnection with Yoruba landscape urbanism, and propose qualitative landscape design algorithms. It is a metanarrative of ES of GI in Yoruba cities towards translating Indigenous Knowledge System (IKS) on ES into design tools. The discussions in the book are situated in urban landscape planning and allied fields by giving consideration to the concept of city and city development, their geographical, spatial and socio-cultural contexts, and achieving a goal of urban planning in the development of eco-friendly cities through local wisdom embedded in IKS. This is possible because the use of local wisdom is still prevalent in landscape management of Yoruba cities and most other Nigerian cities. Urban planning challenges in Nigeria in terms of lack of effective urban governance (no city in Nigeria is self-governed as in the cases of mayoralty system in cities of some other countries), weak legal planning framework, and unavailability of development plans at city and sub-city scales have made issues that have to do with urban development challenging. Yoruba cities occupy Osun, Ondo, Oyo, Lagos, Ogun, and Ekiti States and parts of Kwara, Edo, Delta, and Kogi States in Southwest Nigeria as a single homogenous ethno-linguistic group with a population of 44 million as at 2021. The cities include mega cities like Lagos and Ibadan; big size cities like Ile-Ife, Osogbo, Ado-Ekiti, Akure, Oyo, Abeokuta; and small size cities like Idanre and Oke-Igbo.

World population projection suggests that about two-thirds of humans will dwell in cities by 2030 with the developing world as urbanisation hotspots. Presently, more than half of the world's population lives in urban areas with attendant pressures on land use and the GI in these cities. The literature is replete with the negative effects of these pressures and their skyrocketing impacts on the ES of green spaces in cities. Proof abounds that GI has the capacity to inform design and planning decisions that lead to increased environmental quality and wellbeing of urban dwellers in terms of health and synergetic relationship with the environment where they live, work, and recreate.

Furthermore, ES of GI have been adequately researched in the global north but only few studies have documented them in Africa. Most of the very few studies in Africa are dominated by those carried out on cities in South Africa. This book is also an effort geared towards reducing this imbalance in knowledge generation and sharing by focusing on West Africa with emphasis on a cultural bloc, the Yoruba cities, with peculiar cultural value-system of city planning in time and space. The results of the ethnography in the book generally suggest that the presence and use of GI in the cities fall along the gradient of city cores, intermediate areas, and new sectors. The cultural green spaces are not only located at the city cores which are old areas but demonstrate manifestation of IKS of ES with cultural milieu that transverses religious, economic, social, academic, and political boundaries from pre-colonial times to date. With their cosmological intricacies and metaphysical ES, the cultural green spaces generally referred to as groves are the foundations and the material manifestations of the gods and goddesses in Yoruba antiquity. The intermediate and new sectors of the cities contain English types of parks and gardens with non-spiritual ES much similar to those of the global north cities. These trajectories are detailed in the book.

The book is written in five chapters. Chapter 1 introduces the concepts of ES of GI in cities and their connections with ecological urbanism with focus on explaining the ontology and epistemology standpoints that guided the methodological praxis. Chapter 2 presents an ES survey of GI in selected Yoruba cities. Chapter 3 addresses the ES of homegardens and IKS for human wellbeing. Chapter 4 examines the distinctivenss (and points of confluence) between biophilia and sacredness in ES. Chapter 5 defines approaches to urban landscape design and planning that are eco-cultural by attending to a concluding quest. The quest is on the dialectic puzzle of balancing traditionality with modernity and the means of translating local wisdom into design and planning tools as policy guidelines. On the whole, the book showcases the ES of GI in the sub-region as cosmological, gastronomic, and medicinal green spaces. Written largely from the perspective of landscape hermeneutics, it contains insights into the origin, mechanisms, and sustainability of the IKS. The book promises to be resourceful for urban policy makers, researchers, and practitioners.

Federal University of Technology
Akure, Nigeria

Joseph Adeniran Adedeji

Acknowledgements

The following organisations contributed immensely to the success of this book and are acknowledged for their supports: Alexander von Humboldt (AvH) Foundation, Germany, for sponsoring the fellowship project; Faculty Environment Design Therapy, University of Applied Sciences (Hochschule für Wirtschaft und Umwelt, HfWU), Nuertingen-Geislingen, Germany, for hosting the fellowship; and Federal University of Technology Akure (FUTA), Nigeria, for release during the fellowship.

I thank the following individuals for their much-valued supports toward the success of the book: Prof. Dr. Hans-Christian Pape, AvH Foundation President; Dr. Enno Aufderheide, AvH Foundation Secretary General; Alexandra Justus, my contact person at the AvH Foundation; Prof. Dr. Roman J. M. Lenz, the Dean (and Chair) of Faculty Environment Design Therapy, my host at HfWU; Prof. Dr. Andreas Frey, HfWU Rector; Dr. Ellen Fetzer, Coordinator of the International Masters in Landscape Architecture (IMLA) programme and President, European Council of Landscape Architecture Schools; Prof. Dr. Iris Ramme, Director of the HfWU International Office; Carola Staib, the officer in charge of Visiting Professors at the HfWU International Office; and Corinna Allevato, Miriam Michelsen, Martina Hart, Udo Renner, Jutta Schnell, among many others.

I am highly indebted to Emeritus Professor of Landscape Architecture and Urban Forestry, Prof. Dr. Alan Simson, Leeds School of Art, Leeds Sustainability Institute, Leeds Beckett University (UK); Dr. David Taiwo Adetoyese Oyedemi of the same University; Prof. Joseph Akinlabi Fadamiro, my doctoral supervisor; Prof. Isah Bolaji Kashim, the immediate past Dean, School of Environmental Technology (SET), FUTA; Prof. Yomi Michael Daisiowa Adedeji, the immediate past Head of Architecture Department; Prof. Isaac Olaniyi Aje and Prof. Gabriel Fadairo, the incumbent SET Dean and Head of Architecture Department respectively; the Management of FUTA and the Vice Chancellor, Prof. Adenike Temidayo Oladiji, and immediate past Vice Chancellor, Prof. Joseph Adeola Fuwape and the Deputy Vice Chancellor (Academic), Prof. Deji Rufus Ogunsemi, among many others.

I owe appreciations on this book to the following people: Dr. Olumuyiwa Bayode Adegun, Prof. Ganiyu Oboh, Prof. Jonathan Onyekwelu; Prof. Daniela Perrotti, Chair, Landscape Architecture, University of Louvain, Belgium; leaders and members of the Deeper Life Bible Church, Stuttgart Region of Germany; and Dr. Ezra Gayawan. Springer Nature *Cities and Nature* series editors, Prof. Peter Newman, Prof. Cheryl Desha, Dr. Alessandro Sanches Pereira, and the anonymous reviewers. Thanks to Juliana Pitanguy, the Publishing Editor; Malini Arumugam, the Project Coordinator (Books); Corina van der Giessen, Project Coordinator; and Marion Schneider, Senior Editorial Assistant. I appreciate the following scholars for their expert appraisal of the second revision of the manuscript of this book: Prof. Dr. Magnus Treiber, Institute of Social and Cultural Anthroplology, Ludwig Maximilian University, Munich, Germany; Prof. Oluremi Akinropo Akindele and Dr. Oluwole Philip Daramola, Department of Urban and Regional Planning, Ladoke Akintola University of Technology, Ogbomoso, Nigeria, and Obafemi Awolowo University, Ile-Ife, Nigeria, respectively. I am highly indebted to my wife, Temileye Omotayo Adedeji, and children: John Adeyemo, Joshua Adeyemi, Mercy Adefunke, and Mary Adewumi – Adedeji, for their moral supports.

Contents

About the Author

Joseph Adeniran Adedeji researches the intersections of spatial considerations for the comfortable use of urban open spaces, cultural morphology of cityscapes, and more intensely, landscape hermeneutics of the urban grain in an African context. He holds PhD, MTech, and BTech degrees in Architecture. He is Fellow of Alexander von Humboldt (AvH) Foundation, Germany and Associate Professor in the Department of Architecture, Federal University of Technology, Akure, Nigeria. Joseph was AvH and Research Fellow at University of Applied Sciences (Hochschule für Wirtschaft und Umwelt, HfWU), Nuertingen-Geislingen, Germany. Earlier, he was a Carson Fellow at Rachel Carson Center for Environment and Society, Ludwig Maximilian University, Munich, Germany. Joseph is a full member of the Nigerian Institute of Architects and has full registration of the Architects' Registration Council of Nigeria (ARCON).

List of Abreviations

CES	Cultural Ecosystem Services
CWN	Connectedness With Nature
ES	Ecosystem Services
ECS	Edible City Solutions
F.	Family
GI	Green Infrastucture
HNC	Human-Nature Connections
ICOMOS	International Council on Monuments and Sites
IKS	Indigenous Knowledge System
LCC	Lekki Conservation Centre
MEA	Millennium Ecosystem Assessment
NBS	Nature-Based Solutions
NCF	Nigerian Conservation Fund
UGI	Urban Green Infrastructure
UGBS	Urban Green and Blue Spaces
UNESCO	United Nations Educational, Scientific and Cultural Organisation
UN SDGs	United Nations Sustainable Development Goals
USES	Urban Socio-cultural Ecosystem Services

List of Figures

Chapter 1
Introduction: Ecosystem Services in Yoruba Cities – Towards a Conceptual Framework

Abstract Yoruba cities are located in Southwest Nigeria. The cities have been in existence before the British colonial rule of the region. They manifest distinctive features of ecological urbanism in many ways. In ecological urbanism, cities are interpreted as ecosystems. These are human and nonhuman systems that co-exist to simultaneously provide and access ecosystem services symbiotically. It is argued in ecological urbanism that there is an interface between ecology and urbanism. This interface is explored to redefine cities in terms of metaphor and literally as biological organisms. For this reason, ecological urbanism is an eco-cultural strategy designed to solve the current crisis on urban nature. It is indeed landscape-urbanism-reloaded. It provides a shift of paradigm from defining cities as built forms to redesigning them as ecosystems of nature for human wellbeing. An understanding of this fusion of humans and nonhumans from theoretical perspectives can inform best practices for achieving ecosystem services for human wellbeing. This chapter examines the concepts of these connections to interrogate ecosystem services of green infrastructure in Yoruba cities. The emerging conceptual framework guided the methodological approach of the study meta narrated in this book. This chapter identifies the trajectories of ecological urbanism through seminal works. These include the works of Patrick Geddes, Ian McHarg, Frederick Law Olmsted, Jane Jacobs, Lewis Mumford, Henri Lefebvre, Frank Lloyd Wright, Le Corbusier, Ebenezer Howard, Charles Waldeheim, Anne Whiston Spirn, and Mohsen Mostafavi. The chapter refers to the relevances of ecological urbanism to the 2030 Agenda of the United Nation on Sustainable Development Goals. It then focusses on three quests with reference to Yoruba cities to unpack: (1) the ontological worldviews that are dominant about the *nature* of green infrastructure in Yoruba cities; (2) how these ontological worldviews guide an acceptable epistemological framework for studying ecosystem services of green infrastructure in Yorubacities; (3) and how the methodology of studying ecosystem services of green infrastructure in Yoruba cities emerges from this epistemological paradigm. The emerging conceptual framework for the evaluation of ecosystem services of green infrastructure in Yoruba cities relies on relativist ontology, interpretivist epistemology, and subjectivist methodology. The chapter concludes that the study is not only amenable to this conceptual framework, it is almost a *condicio sine qua non* to studying Yoruba

J. A. Adedeji, *Ecological Urbanism of Yoruba Cities in Nigeria*, Cities and Nature, https://doi.org/10.1007/978-3-031-34688-0_1

ecological urbanism. The next chapter of the book discusses the ecosystem services of specific green and blue spaces in particular Yoruba cities. It aims to understand the mechanisms of the ecosystem performance of the spaces through the conceptual framework developed in this chapter.

Keywords Urban ecosystem services · Urban green infrastructure · Ecological cities · Urban ecosystems · Yoruba urbanism · Yoruba cities · United Nations Sustainable Development Goals

1.1 The Concept of Ecological Urbanism of Yoruba Cities

Current environmental crisis has assumed an alarming dimension (Clark 2020). This calls for a rethink of how cities are defined and designed, including Yoruba cities. Hitherto, the lopsided views of cities in general as built-up spaces are largely responsible for the dearth of green infrastructure in cities (Gavrilidis et al. 2022). Green infrastructure is "an ecosystem or a network of ecosystems with specific parts, needs, functions, and services" (Bianconi et al. 2018, p. 112). It is also "an interconnected network of green space that conserves natural ecosystem values and functions and provides associated benefits to human populations" (Benedict and McMahon 2012, p. 12). The pluralism of green infrastructure makes it "the resilient landscape that supports ecological, economic and human interests by maintaining its integrity and promoting landscape connectivity, while enhancing the quality of life and sense of place of the environment across different landscape boundaries" (Mell 2010, p.225). In this notion, landscape is acceptable as "integrative and visible socio-ecological systems of natural, physical, cultural, social, and aesthetic properties across different spatial-temporal scales" (Landscape Europe 2015, p. 16). Green infrastructure are supposed to provide ecosystem services that meet the peculiar demands of city dwellers in different contexts including their everyday living and endeavours for wellbeing (Perrotti and Iuorio 2019). These definitions suggest that green infrastructure is a major pro-ecological element of ecological urbanism, along with rivers, bays, hills, climate, culture and human history (Mumford 1968).

Ecological urbanism is a functional tool for redefining cities from the perspective of green infrastructure to enable their redesign to achieve sustainable goals through appropriate ecosystem services. It is an eco-cultural strategy whereby urban places are redefined as being composed largely of requisite green spaces 'sprinkled' with essential built forms. Achieving such feat of urban design would require evidence-based approaches. However, there is overtly wrong generalization of such evidences in the western Eurocentrically-dominated literature (Spirn 2014). This calls for reexamination of these perspectives with non-western lens in specific urban contexts. Such reexamination could debunk the over generalization of such concepts as nature-based solutions and biophilic ideologies (Ferreira et al. 2020; Bayulken et al. 2021). The overflogged generalizations fail to recognize local patterns and Indigenous Knowledge Systems (IKS) that are germane to the wellbeing of urban dwellers. Specifically, the peculiar cases of the global south suggest the need for seeking alternative approaches dictated by the people concerned.

In the global south, the Yoruba people form a large single homogenous ethno-linguistic group in Nigeria and Africa. The Yoruba homeland in this British West Africa lies between Longitudes 2°30′ and 6°30′ West, Latitudes 6° and 10° North of the Equator covering about 181,300 square kilometers (Atanda 1996, p.4). They have a population of over 44 million (WPR 2021) in this homelandas shown in Figs. 1.1 and 1.2. They occupy the following states in Southwest Nigeria: Osun, Ondo, Oyo, Lagos, Ogun, Ekiti and parts of Kwara, Edo, Delta and Kogi States as shown in Fig. 1.3. The diasporas in Benin Republic, Brazil, Ghana, United Kingdom, Europe, and Americas are enormous. The Yoruba homeland is an urbanized nation with mega, big, medium, and small size cities. For example, while Lagos and Ibadan fall to the category of mega cities, Ile-Ife, Osogbo, Ado-Ekiti, Akure, Oyo, Abeokuta, are big size cities, Idanre and Oke-Igbo being small size cities. In this vast ethno-geographical space lies the city of Ile-Ife from antiquity. Ile-Ife is at the same time the ancestral home and cradle of Yoruba race and mythological origin of all humans of the world habitat. Throughout the Yoruba land and beyond, Ile-Ife and its sacred urban spaces are highly referred. To show the ecological status of Yoruba cities, Aluko et al. (2008) argue that "the Yorubas have respect for the earth, not only for humans, but also for plants, animals, rocks, burials" (p.149), suggesting a complete urban ecological system. Osogbo is another important Yoruba city. The presence of Osun Grove UNESCO World Heritage Site at Osogbo as archetypal of Yoruba urban

Fig. 1.1 Map of Africa showing the location of Yorubaland. Source: Anderson (2006)

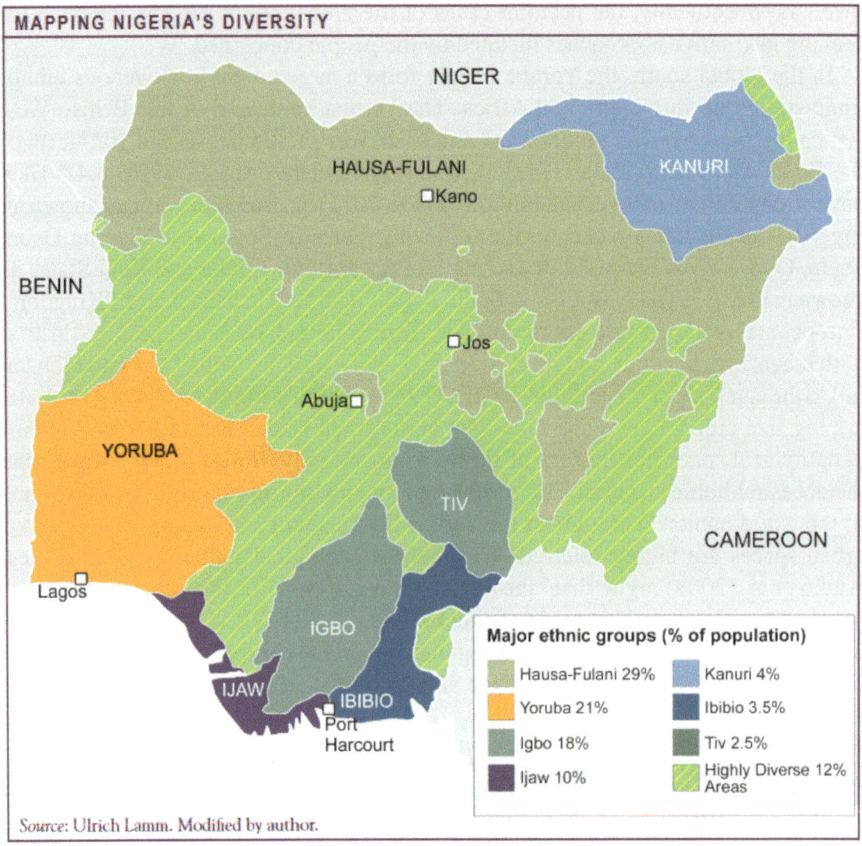

Fig. 1.2 Map of Nigeria showing the spatial distribution of major ethic groups. Source: Kwaja (2011)

groves makes it suitable as a case study of ecosystem services. The birth, growth, and development of Osogbo revolve around this urban grove which has also popularized the city from local through national to international prominence.

The geographical, spatial and socio-cultural contexts of Yoruba cities under investigation require a dissection of the concept of city to assess the ecosystem services of their green infrastructure. City is one of the most difficult categories to conceptualize in urbanism literature. This is because of the divergence of ontological perspectives between the Western and non-Western notions of how a city should look like, what characteristics should embody its identity, and disciplinary perspectives. Cowgill (2004) argues that "it is notoriously difficult to agree on a cross-culturally applicable definition of the city" (p. 526). Gates (2011) puts it plainly that the definition of a city, and therefore the urban, "can vary according to the position of the observer" (p. 2). This immediately presupposes that the idea of city cannot be fixed nor based on *realism*. The city itself is analogous to an ecological

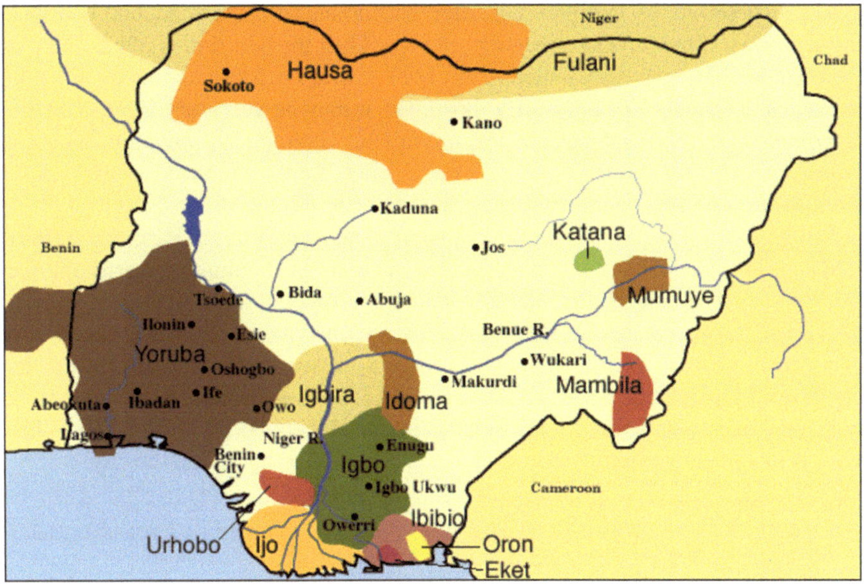

Fig. 1.3 Map of Nigeria showing the location of Yoruba cities. Source: Anderson (2006)

organism that has a lifecycle of birth, growth/development, and can *die* due to decay and has capacity to be resuscitated through urban renewal processes (Jacobs 2016; Adedeji and Fadamiro 2018). The dynamic character and numerous typologies of cities suggest a relativist ontology. In terms of genealogy, historical account of cities suggests that cities started to evolve in the Iron Age between 1300 B.C. to 600 B.C. Since then to date, human civilization has been synonymous with the emergence of cities of differing characteristics from that era through the classical periods of Ancient Greece, Ancient Rome, Persian Empire, and Byzantine Empire, to the Medieval, Early Modern, and Modern eras (Bairoch 1988). Archaeological evidences (Gates 2011) reveal that early cities in human civilization did exist and "sheer number of people do not determine a town, or for that matter a city". For these reasons, "we must see a town or a city as a center, not only of population but of religion, the arts, governance, the military, industry, or commerce" (Hull 1976, p. 388). Importantly, earlier demographic conceptualization of a city has been discovered to be inadequate considering the ancient and influential cities of antiquities that lack the demographic sizes of modern cities. The sociological perspective of the city by Wirth (1938) that "a city may be defined as a relatively large, dense, and permanent settlement of socially heterogeneous individuals" (p. 8) is not specific about demographic characterization of a city.

Lewis Mumford (1895–1990) was another foremost thinker about urbanism and distinguished analyst of the idea of a city. He debunked rather that "it is art, culture, and political purpose, not numbers, that define a city" (Mumford 1961, p. 125). Smith (2005) observes that governance factors are important to the idea of city and its size. Therefore, as human civilization and urbanisation 'onslaughts' continue, the very idea of city and its meanings change. Then the question is, what qualifies the Yoruba urban spaces presented in this book as cities before situating them within ecological urbanism? To demystify this quest, it may be necessary to propose that city ontology of the Yoruba can be legal, symbolic, economical, spatial, metaphoric, or different combinations of these characteristics. Lewis Mumford (1937), in his work, *What is a City*, argues that the city is "a theater of social drama" and everything within it contributes to that drama. In another conceptualization, Park and Burgess (2019, p. 2) contend that a city is an "ecological unit" and "cultural patterns of urban life" (p.viii). They assert that "the city is not an artifact or a residual arrangement" but rather "the city embodies the real nature of human nature. It is an expression of mankind in general" (p.ix). In another sense expressed by Ward (1978, p. 1), "the city is not an artifact or a residual arrangement […]. It is an expression of mankind in general and specifically of the social relations generated by territoriality". This view is in agreement with Park and Burgess (2019) who insist that the city:

> is something more than a congeries of individual men and of social conveniences, streets, buildings, […]; something more, also, than a mere constellation of institutions and administrative devices- courts, hospitals, schools, police, and civil functionaries of various sorts. The city is, rather, a state of mind, a body of customs and traditions, and of the organized attitudes and sentiments that inhere in these customs and are transmitted with this tradition. The city is not, in other words, merely a physical mechanism and an artificial construction. It is involved in the vital processes of the people who compose it; it is a product of nature, and particularly of human nature (p. 1).

These definitions are in agreement with the proposal that Yoruba cities are conceptually legal, symbolic, economical, spatial, and metaphoric entities. A Yoruba city is comparable to a complex dwelling machine with many interwoven parts laced with mechanisms, frictions, operations, functions, actors, and dynamics. Therefore, Yoruba cities are synchronously physical and metaphysical entities, combinations of processes and products, actions and actors, coding and decoding, and indeed a genre of all possibilities. Generally, the complexity of the city makes it amenable to too numerous and sometimes conflicting definitions, criteria, perspectives, agenda, concepts, and modes of enquiry. The city has been defined through demography, infrastructure, administration, politics, culture, economy, society, history, power, geography and from multiple ideologies some of which have toxic layers. It is the document of civilisation and its historical realities.

These multifarious notions of the city are loaded with values of urban places which can be summarized as humanism, symbolism, culturalism, biologism, imageability, governmentality, permanency, spatiality, organism, and economy, among others. These strands, though could be superficially self-conflicting, signify the complexity of the city generally and Yoruba cities in particular. The perception of the city from a humanist ideology would essentially imply making human wellbeing an indispensable goal of urbanism. This is linkable with nature-based solutions and emerging urban environmental crises. It is interwoven with the understanding of the city as an intricate formation of human landscape and urban landscape. Seeing, understanding, framing and accepting the city as a place of "humanist culture" has power that can make humans develop "affection" for nature through cultural practices in cities (Bevir 2021, p. 1). Girard (2016, p. 551) argues that "humanism approach" can contribute to the implementation of different layers of symbioses between "landscape conservation" and opening up of new opportunities in urbanisation processes that can affect the "quality of life" of urban dwellers. The city symbolizes many things and everything. Through perceptual mode, the complimentary symbolic symbiosis between human and nonhuman or more-than-human nature could be sensed to be noticeably exhibited in Yoruba cities and their ecological urbanism. With symbolism, there is differentiation between the city (and its constituents) as the nature 'outside' of humans (nonhuman nature) and nature 'inside' of humans (human nature). Lewis Mumford (1938) in *The Culture of Cities* defines the city from another perspective that is symbolic. To Mumford:

> The city, as one finds it in history, is the point of maximum concentration for the power and culture of a community. It is the place where the diffused rays of many separate beams of life fall into focus, with gains in social effectiveness and significance. The city is the form and symbol of an integrated social relationship; it is the seat of the temple, the market, the hall of justice, the academy of learning. Here in the city the goods of civilisation are multiplied and manifolded; here is where human experience is transformed into viable signs, symbols of conduct, systems of order (1938/2016, p. 104).

In Mumford's thought about the symbolism of the city, it is also:

> A geographic plexus, an economic organization, an institutional process, a theatre of social action, and an aesthetic symbol of collective unity (Mumford 1938, p.185).

The city has also been analyzed through the organismic perspective to understand and profer solutions to its human-induced environmental challenges. Amati (2021) framed a wider discourse around this theme to discuss naturalism as a potent apparatus to unpack city life through diagnostic and prescriptive strategies in medical sciences. Fadamiro and Adedeji (2012) draw parallelism between the city and living organism in term of morphological system. They argue that the morphology of cities is describable by the morphology of biological organisms through biologism. Biological organisms consist of cells with a nucleus, tissue comprising of cells, organ containing tissues, system coalescing a number of organs and the body having a number of systems. This is revealing to explain the working of city mechanisms. In this way, the city is not only a congeries of ecosystems, but a composite living organism. The ecological notion of the city is affirmed by Leong (2012, p. 2) that

> Cities represent the convergence of identities, industries, and ideologies in a dynamic ecosystem […], a city is a living, breathing organism with vital systems […], cities are a cultural reflection of our common humanity in all of its beauty and depravity.

A city is thus a place with perpetual urban spaces filled with human and nonhuman identities and governed by a political system which supports economic and cultural lifestyles. The biocultural identity of Yoruba cities makes them amenable to the biological analogy.

Yoruba cities are fittingly situated in these symbolic descriptions of the city. The valorization of the city as a cultural and biological entity through the notions of culturalism and biologism is not a new remix. The city is synchronously a cultural and biological 'being' through the notion of biocultural diversity. To capture this essence, Maffi (2007, p. 269; 2018) appositely defined biocultural diversity as "the diversity of life in all of its manifestations: biological, cultural, and linguistic, which are interrelated within a complex socio-ecological adaptive system". These diversities are nowhere else more manifest than in the city. The ecological fabrics of the city are therefore culturally framed to embrace all but different notions of nature in different urban places. Yoruba cities are not only synonymous to biological living organisms, but are also the containers for all human and nonhuman living things in urban ecological systems. Together with the non-living components of Yoruba cities, the duos of human and nonhuman systems of the cities inevitably determine and dictate their outlooks termed image ability. Image ability is defined as "a pattern of high continuity with distinctive yet interconnected parts" (Rising 2019, p. 424). In Kelvin Lynch's idea, image ability has three components of structure, identity and meaning (Lynch 1960). It leads to an entire gamut of human responses to the city's ecological spaces and places. Image ability of Yoruba cities leads to the development of clear and unambiguous sense of place, place identity, place perception, place meaning and the attendant attachment of humans to certain places in the cities.

The importance of ecological urbanism as a tool for analyzing and understanding urban patterns is demonstrated by the morphology of these Yoruba cities. The Yoruba cities are ecological milieus where dwellers are human organisms that are not living over but living with non-human nature. The cities lend themselves to learning about ecological urbanism. Conversely, ecological urbanism could explain the cultural mechanisms of human-nonhuman interactions and ways of human living in these cities. The flora and fauna components of the non-human dwellers in these ecological systems (ecosystems) support the wellbeing of their human co-dwellers. The ecology of the Yorubacities therefore presents a complete system of human and other-than-human (or nonhuman as intergeably used and meant as the same thing in this book) elements that are in symbiotic relationships. The urban spaces of these Yoruba cities satisfy the four identifiable prerequisites of ecological urbanism. These are compactness, complexity, efficiency, and stability (Tadi and Manesh 2014), which qualify them for evaluation of urban ecosystem services with the lens of ecological urbanism.

Yoruba cities are characteristically compact. Their compactness results from their close-knit urban form, being produced from peculiar socio-cultural processes and mechanisms. Bascom (1962, p. 699) asserts that "the Yoruba of Nigeria are the most urban of all African peoples of comparable size….and they have large, dense cities which existed prior to European penetration". The cities are also complex. This complexity of urban form is not necessarily morphological. It is about semiotic urban imagery that is produced from cosmological intricacies. For instance, the King's palace is always characteristically central and the four major cardinal routes to the city radiate from the palace according to the cosmological considerations of the traditional gods. The east-facing route is associated with *Sango*, the west-facing route with *Esu*, the north-facing route with *Obatala*, and the south-facing route with *Ogun* (Obateru 2006). The shrines of these gods are located in their respective directions around the central palace. The palace has a complex form and spatial configurations of many courtyards. The courtyards, called *agbala*, are delineated into public, semi-private, and private categories. The public courtyards are accessible to common citizens, especially during corporate city functions like annual idol festivals and celebrations. The semi-private courtyards are accessible to palace chiefs and members of the kingship institution. The private courtyards of the palace are also divided into two groups. The first group of private courtyards is accessible to the king's immediate family members, including wives called *Oloris*, princes and princesses. The second group is accessible to the king only. This may be at annual interval or just once in a life-time for the coronation ceremony.

The functional efficiencics of Yoruba cities are showcased in their perfectly organized urban systems. There are communal open spaces like the King's market square around the central palace, neighbourhood communal open spaces, compound housing with spacious open courtyards, and groves. The king's square doubles as Public Park for communal meetings, traditional religious festivals, and rallies. The big courtyards of the compound housing serve many purposes. These include family events like naming ceremonies and associated rituals, meetings, and settlement of inter-personal and communal quarrels. Others are, collection of rainwater into the impluvium through the compluvium, post-harvest processing and sun-drying of agricultural produce, and traditional religious purposes. The courtyards mostly contain shrines of family idols. There may be series of home gardens around the dwellings. These include kitchen and yard gardens. The waterscapes around or within the cities also provide ecosystem services that are for utility and cosmos considerations. Thus, the entire urban spaces of Yoruba cities are material manifestations of socio-cultural lifestyles conceptually. The social components of the cities showing the hierarchical organization of the kingship institution and the cult systems. The stability of this Yorubaecological urbanism cannot be overemphasized. It has witnessed decades of permanence despite the pressures of urbanisation, civilisation, modernisation, and globalisation.

Invariably, ecological urbanism of Yoruba cities emanates from Yoruba ways of life and cultural domain. This domain could be summarized as Yoruba urbanism. Arguably, Yoruba urbanism is ecological. This ecological framework is a duo that mirrors the harmonious relationship and biomimicry between the heavenly celestial and the earthly terrestrial systems. In the heavenly celestial territory, it is believed in Yoruba cosmology that a divinely-ordered relationship exists between *Olodumare* (the Yoruba name for God) and certain 16 elders. These elders are believed to be assigned by *Olodumare* to follow Oduduwa (the projenitor of the Yoruba race) to create the world at Ile-Ife. Four of these 16 elders are believed to be cardinal. For this reason, they are the gods that are worshiped in the shrines along the four major cardinal points. The other twelve elders include the god of divination called *Ifa* (*Orunmila*), *Oluorogbo* (the messenger between humans and gods), and Oramfe, among others. To buttress the unity in this celestial system, the belief in Yoruba cosmology is that *Olodumare* sent Oduduwa to create the world at Ile-Ife along with these deified elders. They are believed to have various functions in the creation narrative and sustain the effective functioning of the created world and its urban spaces (Gordon 1979; Falola and Genova 2005; Fandrich 2007).

As exemplified with Yoruba cities, ecological urbanism seeks and argues for an ecological approach to remaking existing cities by integrating ecology into their urban spaces through design and redefining future cities as eco-spaces. It brings a paradigm shift in understanding of cities from built forms (where nature has to be 'designed-out') to an urban-ecological framework (where the city is understood as an ecological entity). Humans and non-humans co-dwell in a natural city environment with balanced ecosystem. Instead of 'seeing' the city as built (artificial) ecosystem of human dwellings distinguished from unbuilt natural ecosystem of non-humans, ecological urbanism in Yoruba cities unifies both artificial and natural ecosystems into the city. Therefore, Yoruba cities are complete ecological environments of all living and non-living things, physically visible or otherwise. In these cities, ecological urbanism unites the biotic, abiotic, and socio-cultural systems as a single and sustainable urban entity. The biotic (living) components include all categories of flora and fauna, including humans. The abiotic (non-living) components include all elements of the lithosphere, hydrosphere, and atmosphere. The socio-cultural components are essentially the human systems of kinship, belief-systems, language, religion, economy, and all other value-systems. Thus, the components of ecological urbanism in Yoruba cities include and cater for all categories of flora, fauna, including all socioeconomic and socio-cultural categories of humans. These are the rich, poor, young, middle-aged, old, educated, uneducated, and religious belief systems (traditional and modern). Others are the physically-challenged, able-bodied, ruling elites, peasants, and all socio-economic strata. All the non-human categories in the aquatic, terrestrial, and arboreal habitats are parts of these Yoruba urban systems, called ecosystems.

This book pushes for the realization of the 2030 Agenda of the United Nation Sustainable Development Goals (UN 2019) in Yoruba cities in particular and global south cities in general. This requires developing a conceptual framework for understanding the ecosystem services of green infrastructure in these cities. The overarching goal is to develop strategies for actualizing the urban components of goal 2 and goals 11 to 15 from ecological perspectives of urban design in Yoruba cities using systems' approach (Barbier and Burgess 2017). Ecological urbanism is critical to achieving goal 2.4 on the maintenance of ecosystems and goal 2.5 on associated traditional knowledge through sustainable urban design. Similarly, goal 6.6 on the conservation and restoration of aquatic landscapes is central to ecological urbanism in Yoruba cities. In these genres, Indigenous Knowledge System (IKS) has been identified as a foundational component of biological and ecological diversities (Kosoe et al. 2020). Therefore, ecological urbanism of Yoruba cities is an ecological urban design that guides the identification and integration of what nature can do for city dwellers through IKS and vice versa. In terms of energy flow, it redefines the city from the perspective of a metabolism engine. The analogous metabolic engine inputs energy and other resources and outputs wastes to an eco-cultural system which is mostly neglected. In view of this, ecological urbanism has capacity to explain energy circulation for sustainable energy systems in Yoruba cities as required in goal 7.1 of the United Nation Sustainable Development Goals (UN 2019). The participatory sustainable human settlement planning (Nilsson et al.

2021) required for the achievement of goal 11.3 can only be a mirage without ecological urbanism as exemplified for Yoruba cities in this book. Achieving these goals would require ecologically-inclusive cities where the hard distinction between bionetwork and urbanity has been dissolved through ecological urbanism. In these bionetworks of ecosystems termed cities, the non-human (and more-than-human generally) and human nature mutually co-exist and are co-beneficial to one another.

Furthermore, the integration of accessible and socially-inclusive green spaces required in goal 11.7 is a task for ecological urban design of cities and has capacity to explain the ecosystem services of green infrastructure in Yoruba cities. This implies that requisite private and public green infrastructure that meet all the needs of provisioning, supporting, regulating, and cultural ecosystem services for the well-being of dwellers in Yoruba cities are not negotiable. These cannot be jettisoned for metabolic urbanism but rather integrated with it. Even though what is meant by 'nature' is not defined and may mostly refer to only material manifestation in goal 12.2, its focus on the sustainable use of natural resources is paramount for assessing ecosystem services of green infrastructure in Yoruba cities. Application of this goal in practice would require specificities of interpretation to transform it into practical realities in the urban design of Yoruba cities. This is a task for ecological urbanism to identify and apply the different contextualization of nature in diverse places, including Yoruba cities. Nature could refer to organic nature of the pristine environment, capitalist nature of the cultivated agricultural spaces, or techno-nature of biotechnology products (Escobar 1999), and essentially, the 'spiritually-cultivated' mystic nature in Yoruba cities. The health impacts of chemicals and waste management strategies on human and non-human dwellers in Yoruba cities are parts of the requirements for environmental sustainability in goal 12.4 of the United Nation SDG. They would require proactive measures through understanding of the cities with the lens of ecological urbanism. The significance of ecological urbanism to achieving goal 12.b on sustainable tourism that promotes indigenous culture cannot be overemphasized in Yoruba cities. There is no doubt that ecological urbanism is a crucial urban design strategy for mitigating the negative effects of climate change in Yoruba cities as required in goal 13.b of the SDG. Assessment of coastal cities in Yoruba land with the lens of ecological urbanism would benefit maximally through the enhancement of coastal ecosystem resilience to achieve goal 14.2 and their conservation required in goal 14.5 of the SDG (Neumann et al. 2017; Reimer et al. 2021).

The conservation and sustainability of terrestrial and inland freshwater ecosystems and services including urban green spaces, groves and waterscapes in Yoruba cities are ecologically indispensable. These align with the stipulations in goal 15.1 of the SDG and the cities stand to benefit maximally from ecological urbanism in their (re)design. Goal 15.4 of the SDG is on sustainable mountain ecosystems and their biodiversity, including the prevention of extinction of threatened species of flora and fauna in urban wildlife ecosystems (Opoku 2019). The conservation of these unique ecosystems in Yoruba cities are achievable through ecological urbanism. Other United Nation SDG that are ecologically peculiar to Yoruba cities include the sustainable conservation of ecosystem biodiversity values in goal 15.9 and capacity enhancement of local communities for sustainable livelihood in goal 15c.

The foregoing suggests that ecological urbanism is poised to be an overarching urban phenomenon and a major agent of sustainable cities of today and the future in Yoruba land. It counterbalances the emerging multiple ecological dislocations in the current urban grains engendered by uncoordinated urbanisation in many parts of the world, including Yoruba land. In view of the huge socio-cultural human connections to these detates, ecological urbanism is inherently interwoven with socio-cultural ecosystem services in human-inhabited urban spaces of Yoruba cities. Unfortunately, there is paucity of knowledge on these ecological parameters for sustainable urbanism in sub-Saharan Africa at large and Southwest Nigeria in particular (Du Toit et al. 2018). This calls for a nuanced understanding of the connection between ecological urbanism and urban ecosystem services in general. Specifically, the link between socio-cultural ecosystem services and socio-cultural diversities calls for assessment of ecological urbanism of Yoruba cities.

1.2 Urban Socio-cultural Ecosystem Services of Yoruba Cities

People in different cities and urbanized parts of the world, including Yoruba land, relate with the natural environment in different ecological ways (Adedeji 2020; Cocks and Shackleton 2020a, b). These relationships result in human-environment connections and associated ecosystem services. These categories of ecosystem services are generally referred to as socio-cultural ecosystem services. They consist of both material and non-material benefits that humans derive from urban ecosystems and natural environments in particular and contribute to quality of life (Bullock et al. 2018). These ecosystem services are generally based on indigenous knowledge system (IKS) and have capacity to enhance wellbeing of dwellers in Yoruba cities through physical and metaphysical strategies (Adedeji 2020). Urban Socio-cultural Ecosystem Services (USES) has been linked to successful social resilience in cities through the concept of Edible City Solutions (ECS) to leverage on the knowledge of Nature-Based Solutions (NBS) (Wilk et al. 2021). While ECS include practises of "urban farming combined with closed loop systems for sustainable water, nutrient, and waste management", which are inherent in ecological urbanism, NBS generally build upon the established input of well-administered and sundry ecosystems to enhance human resilience (Säumel et al. 2019, p. 1). Potschin et al. (2016, p. 1) argue that NBS is "an umbrella term for all related applications of ecosystem services", ecosystem strategies, ecosystem-based methods, "natural capital and lessons from nature". USES involve the social and cultural practices of using nature in Yoruba cities to achieve wellbeing while NBS and ECS are not only closely interlaced with it but USES are achievable through the duos. In particular, it has been discovered that research into cultural ecosystem services (CES) of urban green infrastructure (UGI) needs to put social and ecological factors into consideration to achieve liveability of cities (Dickinson and Hobbs 2017). This makes the definition of CES by MEA (2005, p. 39) as "nonmaterial benefits people obtain from ecosystems through spiritual enrichment, cognitive development, reflection, and aesthetic experiences" very appropriate for examining the ecosystem services of green infrastructure in Yoruba cities. With this nonmaterial perspective in understanding the value of CES and incorporating a social component, Wallace (2007) submits that socio-cultural services offer spiritual and philosophical satisfaction, access to companions and being loved, recreation and leisure, and meaningful attachment to natural environments. These perspectives are crucial for developing a conceptual framework towards understanding ecological urbanism of Yoruba cities.

The literature provides deep insights into the significance of USES in supporting the bodily and mental wellbeing of urban dwellers that are applicable to Yoruba cities. The social context of USES in Yoruba cities also offers converging perceptions for their different social groups. There are only scanty shifts in the ways and manners ecological responses are socio-culturally co-produced among the social groups in view of common cultural index. Like elsewhere, there are few divergencies to the mainstream CES in Yoruba cities, being largely based on cultural morphologies of the cityscapes (Fadamiro and Adedeji 2012). For instance, elsewhere, young people dwelling in city cores tend to prefer CES that enhance social interactions whereas old people in periurban spaces favour ecosystem services that connect them to natural experiences (Riechers et al. 2018). Kim and Son (2021) reveal that these connections could also produce social-cultural dynamic flows, both spatially and temporally. Dynamic flows in CES have also been discovered along urban-periburban gradient (Riechers et al. 2018). Many urban areas with high population densities like cores experience less CES because of the dearth of required quality and quantity of UGI compared with periurban areas. In the periurban areas, city dwellers have more access to UGI with USES including recreation, education, aesthetic appreciation, and overall awareness of nature (Riechers et al. 2018). These services are also linked with cultural diversities along social parameters and gradients like age, education, and income level, for which this book is set to address. The CES diversities have been confirmed to be linked with spatial and temporal characteristics in urban agglomerations, including their demand and supply forces, as they impact human wellbeing (Xiao et al. 2017). In scenarios where urban residents are not in close proximities to public UGI like parks and gardens to access CES, they turn to their residential green spaces in the immediate surroundings to effectively compensate for the lack (Mao et al. 2020). Consequent upon the foregoing, the interrelation of USES, NBS, IKS, CES, and UGI has capacity to contribute to the development of a conceptual framework for demystifying the ecological urbanism of Yoruba cities.

1.3 Philosophies of a Conceptual Framework for Yoruba Ecological Urbanism

The goal of this section is to elucidate the philosophies that guided the development of a conceptual framework for assessing the ecosystem services of green infrastructure in Yoruba cities to demystify their ecological urbanism. The resulting conceptual framework is based on relevant theories on ecosystem services of green infrastructure. Theories are generally describable as systematic arrangements of thoughts for the purpose of examining and explaining naturally occurring phenomena and realities. The purpose of theories is to render accounts that are intellectually acceptable through appropriate justifications. They are scholarly intelligent explanations of procedures for determining the what, where, when, who, and overall contextual considerations of social and natural scientific endeavours. In discussing urban theories, Harding and Blokland (2014) posit that theories are "the outcomes of reflection that sharpen [...] understanding of the world" (p.12). Through theoretical analyses, ecosystem services of green infrastructure could be understood. Metanarration is a tool for stimulating this understanding by showcasing ecological urbanism as eco-cultural lifestyles of oneness with nature of the Yoruba people. The objective of the metanarration is to draw out lessons that modern city-dwellers could learn from the urban ecosystem services in the cities of this distinctive ethnic group. The overall goal of these ecosystem services is the wellbeing of urban dwellers. In addition to enhancing the wellbeing of urban dwellers, understanding the services can guide strategies for pursuing the goal of sustainability of urban ecosystems through Indigenous Knowledge System.

The achievement of these targets would need to be based on appropriate philosophical paradigms in terms of ontology, epistemology, and methodology. These would guide the tracing of the position and actions of humans as a major actor and beneficiary of urban ecosystems services. Through appropriate conceptualization of multiple layers of interwoven relationships in, and reactions to, the urban conundrum, the philosophical underpinnings would validate the resultant evidence-based urban design strategies for Yoruba cities. Urban design is the architecture of cities, cities being big buildings and buildings being small cities. Architects as urban designers have significant roles to play in giving sustainable forms to these cities. This would require understanding the symbiotic relationships between humans, non-humans, and more-than-humans generally and translating them into apposite urban genres through this ecological knowledge. This perspective aligns with the argument of Erixon Aalto et al. (2018) that "the goal of ecological urbanism is to discover systems of nature and reproduce them as urban form" (p.26) as physical manifestation of the ways of life of city dwellers. Therefore, if urbanism is classically "a way of life" (Wirth 1938, p. 1), and way of life is 'culture', then cultures should be analogous to cities. But then, does that mean there is no culture outside cities? Of course no, but there are also 'cultures of cities' and the analogous 'cities of cultures'. The architect is therefore expected to be a form-giving agent to the city-culture homogeny in an ecological manner. Consequently, conceptualizing ecosystem services of green infrastructure of Yoruba cities to study their ecological urbanism would depend on particularities and not on universalisation of the notion of nature (Adedeji 2020; Cocks and Shackleton 2020a, b). For this reason, universalisation of urbanism theories may be counterproductive due to cultural diversities. Most urbanism theories have Western origins and conceptualization and may only be tangentially applicable to non-Western settings like Yoruba cities. Therefore, the concept of what constitutes urban places may not be universal. The intent here is not to parade all the theories about the concept of ecological urbanism but to identify and synthesize critical thoughts that have bearing on formulation of framework for ecosystem services of green infrastructure in Yoruba cities. This approach has capacity to lead to the understanding of the lessons that could be learnt from Yoruba Nation on ecological urbanism and how ecological urbanism discourse can contribute in practical ways to the sustainability of Yoruba urbanism.

The ecological movements on urban discourses have a longterm tradition that is traceable to the seminal works of many ground-breaking thinkers on urbanism. These include Patrick Geddes (Geddes et al. 2021), Ian McHarg (1969), Frederick Law Olmsted (1870), Jane Jacobs (2016), and Lewis Mumford (1938, 2016). Others are Henri Lefebvre (Lefebvre et al. 1996), Frank Lloyd Wright (1971), Le Corbusier (1980), Ebenezer Howard (1946), Anne Whiston Spirn (1984), and Waldheim (2006). Mostafavi (Mostafavi and Doherty 2010) of Harvard Graduate School of design more recently 'crystalized' the thoughts as 'ecological urbanism'. Detail review of this developmental trajectories has been provided by Anne Whiston Spirn (Spirn 2014) and will therefore not be necessary here. With these antecedents to ecological urbanism, the urbanism of Yoruba cities is arguably ecological. Ecological urbanism provides a vision of the urban where the landscape is the basic building block of cities and not architecture. It offers urban spaces that are fully spread and controlled by a system or systems of nature housed in the landscapes. Ecological urbanism provides a wider symbiosis of landscape and architecture that even include what Padoa-Schioppa (2017) describes as "ecology of the mind" (p. 230). This ecology of mind is essentially human and akin to the 'heart' that controls ecological urbanism. These descriptions are ecologically axiomatic of cities in Yoruba urbanism.

While the retinue of constantly emerging urbanisms (new urbanism, landscape urbanism, sustainable urbanism, everyday urbanism, utopian urbanism, etc) appears rhetorical, ecological urbanism foregrounds a triumph of nature in Yoruba cities. There is a triumph of nature in Yoruba urbanism over the artificiality of cities. In line with this triumph, Spirn (2014) advances that the natural entities of urban landscapes (including soil, water, wind, and light) play specific roles. Therefore the city should be understood as being part of nature and nature part of the city. Steiner (2011) is another critical thinker on urbanism who proposes that a full integration of the knowledge of urban ecology into ecological urbanism would produce better results. He argues that:

> urban ecology research indicates what should be obvious: people interact with other humans and with other species as well as their built and natural environments. The city is a human-dominated ecosystem. [.....] Designing with nature can improve the quality of cities for people, plants, and animals. In doing so, ecosystem services can be enhanced (Steiner 2011, p. 336).

Accordingly, he proposed landscape ecological urbanism as a robust synthesis to understanding how cities function. It weaves together urban environmental and human ecologies. These concepts are applicable for understanding ecological urbanism of Yoruba cities.

Tracing the trajectory of ecological urbanism from a perspective that is different from that of Spirn (2014) is important. For instance, Hagan (2014) has questioned the 'newness' of Ecological Urbanism (see Hagan 2014, for a detail discussion of the origins of *Ecological Urbanism*). Susannah Hagan opines that "Ecological Urbanism is simply a late entry into a line of ecological subfields that engage literally or figuratively with the built environment" (p.5). She questioned if "ecological urbanism [*is*] simply a new term rather than a new theory and practice?" (Hagan 2014, p.8). While this position appears extreme, it suggests need for a second look at these trajectories. It may be normal to be guided by objectivity to emphasize that identifying a naturally existing phenomenon and rebranding it for new conceptual application is quite novel. Therefore, ecological urbanism generally needs to be identified as a unique category in understanding city processes and outcomes in Yoruba cities. All along the history of human civilization that cities have been in existence, it could be argued that ecological urbanism has been in existence, starting from the Edenic Garden (Till 2001). There is the need to recall the characteristics of urban places that showcase the principles of Ecological Urbanism and synthesize them for modern practical application to urban design of Yoruba cities. The work of Mohsen Mostafavi and Gareth Doherty appears to have successfully pursued this goal. They argue that ecological urbanism "has the capacity to incorporate the inherent conflictual conditions between ecology and urbanism" (Mostafavi and Doherty 2010, p. 17). Gregory Marinic provides an erudite review of this field in the unique edited volume, *Ecological urbanism*. He argues that Ecological Urbanism is:

> a comprehensive global range of urban interventions, strategies, and proposals that frame a radically interdisciplinary vision of future sustainability [.....*whose*] postulations renounce object-based urbanism by reconsidering the city as an inherently complex ecology of places, people, and processes (Marinic 2012, p.105).

The volume showcases the engagements of diversities of places, people, and processes in the trends and ingenious applications and practices of ecological urbanism towards achieving asustainable world. Sustainability of the world suggests that the earth is a single mega-ecosystem comprising many sub-ecosystems called cities. The excellence of the sustainable urban design strategies and their diversities in the volume comes along with a challenge to the intellectual community to engage more with different peculiar places like Yoruba cities on ecological urbanism.

This book contributes to the knowledge based on landscape urbanism from a non-Western perspective of the Yoruba cities in alignment with the strategy of Mostafavi and Doherty (2010). The fittingness of Yoruba cities to conceptualization of ecological urbanism comes at two complementary levels: bottom-up and top-down. A bottom-up approach to urbanism fits Yoruba cities as a form of cultural urbanism that is produced outside orthodox legal and regulatory frameworks which often results in innovative and ingenious solutions to urban living (Mostafavi 2010). Many of the Yoruba cities predate British colonial influences and therefore were not planned in European ways at the cores. The cities were rather planned through cultural providence of the traditional Yoruba pantheon. This accounts for the types of city layout, open spaces, urban groves, and green infrastructure in these cities. At the top-down end, the influence of colonial city planning remains evident outside the core areas of Yoruba cities as post-colonial footprints.

In the context of urban planning, there is the need to understand the morphology of Yoruba cities that accommodate these green infrastructure. Yoruba cities are mainly homogenous in nature and they have been mainly characterised with evolvement of three basic developmental zones. These are the traditional zone or core area, the transition zone, and the suburban zone. The homogeneity in the morphology of Yoruba cities can be attributable to some factors of development that are common to the cities. Among the factors is the common culture of Yoruba people that is reflected in the spatial arrangement of their cities. Also, the cities have common urban planning and development history of urban development because they have been governed together as a political entity and under common urban planning legislation. For instance, the development of cities in Yoruba region has been largely influenced by the Town and Country Planning Law Cap 123 of the Western Region (comprising the Yorubaland) which replaced the 1946 Town and Country Planning Ordinance after independence in 1960.

The traditional core zone is those developed in the precolonial era, the transition zone comprises the developed areas in the colonial period while the suburban zone consists of areas developed after the independence of Nigeria. Each of the zones is homogenous in terms of physical layout, ethnic composition, socioeconomic status and environmental amenities such as green infrastructure. Thus, access to infrastructure in each zone is premised on their unifying attributes in terms of building layouts, location, type of buildings, and commercial activities which reflect the socioeconomic and cultural attributes of the residents. These developmental zones have their unifying attributes which include the age of buildings, arrangement of buildings or layout pattern, locational attributes and available infrastructural facilities.

(i) **Core area or Traditional zone:** This zone of the town is mainly occupied by indigenes and the buildings in this zone were built before the colonial period. Buildings in this zone are usually closely built together, with footpaths as the main means of accessibility. The houses in this area are the traditional housing forms dominated by the courtyard multi-dwelling buildings that automatically include green courtyards for multiple ecosystem services mainly of traditional courtyard system. Most of the houses in the zone are not habitable as they are usually old, dilapidated and lacking household infrastructure such as water supply, sanitation, solid waste collection system and domestic energy supply, among others. However, the area accommodates the green infrastructure of traditional importance.

(ii) **Intermediate or Transitional zone:** Houses in the transition zone were built during the colonial era. This zone is attributed to both indigenes and non-indigenes employed in the formal sector of the town. As this zone developed mainly during the colonial period, they enjoy fair level of road accessibility and moderate level of accessibility to green infrastructure. In the same vein, they accommodate the green infrastructure that are formally developed for recreational purposes rather than those that are traditionally-oriented in the core area.

(iii) **New or Suburban zone:** Areas in the suburban zone are attributed to high income earners with well laid buildings. The area is located at the outskirt of the town on approved private residential layouts. Its area is more cosmopolitan than the intermediate or transition zone and comprises residential buildings of modern standard with a reasonable level of adequate provision of green infrastructure compared with the previous two zones.

Furthermore, conceptual frameworks are ways of studying phenomena, places, objects, and things. Their importance in urban enquires of Yoruba cities cannot be overemphasized. They provide the conceptual insights into the processes and products of ecological urbanism by illuminating basic concepts of naturally occurring circumstances and events in Yoruba cities. Pickett et al. (2010) understand theories as systems of conceptual constructs. According to Spirn (2005) ecological urbanism unites the theory and practice of urban design. Its conceptual underpinnings are essentially two-pronged harmonized together, one from ecology and the other from urbanism. A near term to ecological urbanism called "eco-city" was coined for the first time by Register (1987) and offered stimulating guidelines for the production of cities with ecological qualities. Sir Arthur George Tansley (1935, p. 229) coined the word "ecosystem" in 1935 to better describe a combination of vegetation and organisms as basic units of nature that have the abilities for self-regulation and self-organisation. Ecosystem has since replaced 'organism' earlier coined by Clements (Golley 1991). Nature is hereby understood as a mega ecological system embodied in green infrastructure. Ecosystem services are hinged on the laws of nature and ecological systems have been generally grouped into three classes of biotic, abiotic, and anthropogenic (Morris et al. 2020).

In Yoruba cities like elsewhere, the concepts of ecology and urbanism are inextricably linked in a complimentary connection to produce applicable guidelines as ecological urbanism. Mostafavi and Doherty (2016) call this the principle of interaction in analyzing the Latin American ecological urbanism of Chile arguing that ecology is basically "the interactions of organisms and the environment" (p.10). This has been christened the concept of 'ecosophy' where a balance is achieved between the city and all its human and nonhuman dwellers through ecological consciousness of deep ecology. Deep ecology is a shift from the hitherto dominant Western worldview of techno-nature to a materialist-spiritualist scientific concept. In this concept, humans are identified and treated as inherent part of Nature and not a separate entity outside it. Accordingly, Kilian Jörg (2019, p. 33) "the ecological reason does not reflect on nature as an external element to it, but sees itself as an immanent part of nature". This proposition calls for identifying and examining how ecological urbanism is conceptualized in different cultural settings like Yoruba cities. It includes how such conceptual *modus operandi* conditions human thoughts in the light of ecological realities. Ainscough et al. (2018) argue that ontological and epistemological bases for particular methodologies are critical to conceptualizing ecosystem services in view of divergence of opinions on the concept. In view of this, three questions are posed on Yoruba cities to guide the development of a conceptual framework for the ecosystem services of their green infrastructure: (1) what are the dominant ontological worldviews about the *nature* of green infrastructure that are applicable to Yoruba cities? (2) in what ways can these ontological positions guide an acceptable epistemological way of studying ecosystem services of green infrastructure in Yoruba cities? (3) and how can an acceptable epistemology guide the choice of method for identifying the ecosystem services of green infrastructure in Yoruba cities. The reason for this premise is based upon multiple understandings on ecosystem services that are hinged on diverse conceptualizations since the concept of ecosystem services is nebulous (Ainscough et al. 2018; Schröter et al. 2014). Therefore, no matter the level of divergences in the literature among multidisciplinary scientists on the amorphous nature of the concept of ecosystem services, the proposed quests are potent to ensuring internal consistency required for knowledge distilling and dispensing in this book.

The sections that follow deal with the literature on the theories and concepts of ecosystem services, green infrastructure, and indigenous knowledge system. The goal is to establish a consistent conceptual framework that is based on the three concepts for analyzing the case studies of Yoruba cities in subsequent chapters. This would enable the understanding of how the indigenous knowledge system of the Yoruba on ecosystem services of green infrastructure could inform evidence-based design strategies for eco-cultural cities.

1.3.1 Ontologies of Green Infrastructure and Ecosystem Services in Yoruba Cities

To attempt answering the first question, ontology has to do with the nature of truth. It guides the understanding of the nature of realities and how truth could be recognized when discovered or it emerges incidentally. The concern here has to do with the truth about realities of urban green infrastructure, their ecosystem services, and Yoruba cities. Traditional philosophy frames much of the literature on ecological urbanism and their antecedents. Two general groups have been identified in this tradition: the realist ontology (*realism*) and the relativist ontology (*relativism*). The proponent followers of realism are called realists while those of relativism are called relativists. Realist ontology is a group with many branches that proposes that there is one and only one truth about any object, subject, place, or thing. Realism implies that the truth is essentially a reality and that this reality exists inside (and not outside) that truth, or the object of the truth under investigation. It could be in the form of naïve realism, structural realism, or critical realism (Danermark et al. 2019). Naïve realism posits that the only one reality about existence of only one truth can only be understood by employing only apposite procedures. Structural realism insists that the only one reality/truth about an object/matter can only be defined by logical principles whose form cannot be made explicit. Critical realism postulates that reality/truth can only be obtained by extensive rigorous investigations (Danermark et al. 2019; Porpora 2018; Hoddy 2019). All these paradigms are considered inadequate for developing a conceptual framework for assessing the ecosystem services of green infrastructure in Yoruba cities because of their rigidity on insistence of only one monolithic truth. Their application would mean that the reality/truth of green infrastructure, ecosystem services, and Yoruba cities cannot be pluralized. Whereas, all the three constructs (green infrastructure, ecosystem services, Yoruba cities) have complex multifarious layers of truth. The complexities are compounded by biodiversities that are peculiar to temporal realities. These dynamisms make realism unsuitable as ontological foundation for assessing and legitimizing the knowledge about the truth of ecological urbanism of Yoruba cities.

Relativism is an ontological position which proposes that the truth about any object or subject is relative to the perspective, places, and people under investigation. It triumphs as a multi-dimensional search-engine for truth and posits that truth is actually outside the object or subject of investigation (Yoruba cities) and is therefore relative. Essentially, relativism argues for and embraces multiple realities. Relativism ontology parades two paradigms: bounded relativism and relativism (Moore 2009). Bounded relativism sets out time and space boundaries within which truth is mentally created within defined cultural, cognitive, moral, social, and many other milieu. Relativism posits that there are multiple mental constructions of truths as realities (Moon and Blackman 2014; Jones 2002). A good illustration is the assessment of the temperature and indoor thermal comfort in a lecture room. Without having to hold any scientific thermometer, all lecture participants who are in relative state of health in their body, soul, and spirit are most likely to confirm almost the same levels of hotness or coldness of the room temperature based on their indoor thermal comfort. On the other hand, extraneous but few individuals in the same room will confirm a different experience that could suggest that they possess a high degree of deviations in the conditions of their states of health in their body, soul, and spirit. Such knowledge is thus diagnostically potent to enable suitable prognosis for amelioration. These individual differences that are implicitly and explicitly accommodated in relativism makes it suitable for studying individual diversities in the ecological urbanism of Yoruba cities.

Yoruba cities have diversities of green infrastructure in diverse urban settings and their ecosystem services are culturally diverse from those of the Western cities. These suggest that the ontological identity of this study is unmistakeably relative. Humans are diverse and the Yoruba people in particular display a socio-spatial dispersion of a united but varied people. This relativism position is also supported by the ecological (biological) and urbanism (cultural ways of life) diversities in human relationship with the biosphere (Essono 2019). There appears to be no better ontological position that can fit this discourse since its aim aligns with relativism on people's belief. Relativism "seek[s] to *explain* what causes beliefs and action and what those beliefs and actions cause" as argued by anthropologist Bernard (2018, p.3) who also argues that this philosophy is practically *humanistic*. In this way, the study is set to demystify the mechanisms driving the ecological relationship between urban green infrastructure and Yoruba people in their 'natural' context, and therefore hermeneutical. In the course of interpreting and making sense of the data on ecosystem services of green infrastructure in Yoruba cities, allusions are made to complex metaphysical rationalizations. The rationalizations are necessitated by the cosmological complexes in Yoruba urbanism. Therefore, metaphysical rationalization are required in some instances. They are "explanations of phenomena by any nonmaterial force, such as mind or spirit or a deity- things that, by definition, cannot be investigated by the methods of science" (ibid, p. 5–6). The indigenous Yoruba people sometimes operate with mystical insights.

Through relativist ontology, there is the need to make sense of the concepts of green infrastructure, ecosystem services, and Yoruba cities. This is to intellectualize the notions of their usage in the book since these concepts bear multitude of meanings to different people in different contexts. Therefore, green infrastructure of Yoruba cities is describable as a category of landscapes in different layers and contexts that binds an ecological relationship between Yoruba people and their urban environments for overall living, survival, resilience, and adaptable symbiotic lifestyles. Yoruba people are inalienably connected to, supported, and sustained by green infrastructure. In the US, UK, Europe, and Australia, the applications of the meaning of green infrastructure have been guided by a number of principles gleaned from the academic and practice literature. These include greenness of spaces, connectivity of green spaces, their multipurpose services, enhancement of ecological network, adaptation to the effects of climate change, and usefulness for sustainable landscape management regimes (Davies et al. 2006; Gill et al. 2007; Mell 2013, 2017). Inevitably, green infrastructure has to do with "the promotion of social, economic and environmental benefits within an integrated approach to planning that enables different stakeholders to shape the ways that they develop and manage the landscape" (Mell2017, p. 135). The two general components of green infrastructure in Yoruba cities are vegetative plants and water bodies. Plants in Yoruba cities include trees, shrubs, grasses, and climbers. Elsewhere, varieties and combinations of these components can be found as woodlands, forests, parks, gardens, playgrounds, and green fields in cities.

In Yoruba cities, there are rich combinations of different green infrastructure components. These include cultural landscapes, sacred rivers and streams, home gardens, kitchen gardens, palace gardens, sacred groves, public parks and gardens, and green yards (in front, sides, and back of houses). Others are green public squares, green courtyards, green shrines, and rich milieu of streets and communal squares sheltered with tropical rainforest trees. They are seen as material and non-material entities in the biological and cultural relationships of the Yoruba people with urban nature (Adedeji 2020). Therefore, throughout this book, green infrastructure in Yoruba cities would mean the material and nonmaterial manifestations of floral elements of the landscape including all plant categories. The plants could be singly occurring or in group or combination of groups, terrestrial and aquatic habitats, both naturally occurring and cultivated in private and public urban domains. They are nonhuman layers of the urban ecosystem that have social, cultural, economic, environmental, biological, and multiple ecosystem service functions for the total environment and indigenous Yoruba people.

Ontological understanding of ecosystem services of green infrastructure in cities is largely conceptualizedby the Millennium Ecosystem Assessment, (MEA 2005) into four groups, namely provisioning, supporting, regulating, and cultural services. However, the ecosystem services research community has constantly debated the consensus definition of the concept. There are oscillations and trying to distinguish among services, benefits, and goods (Kadykalo et al. 2020), implying the pluralities of the conceptualization of ecosystemservicesin the literature. In the case of green infrastructure, provisioning services in Yoruba cities are the direct material resources that are derived by the indigenous Yoruba people from the urban ecosystems for survival. Supporting services are those benefits derived from green infrastructure that ensure the continued existence of the ecosystem. Regulating services are means of ensuring balance in the ecological processes in the urban ecosystems of Yoruba cities. Cultural services are non-material benefits Yoruba people receive from ecosystems and its green infrastructure components. These four ecosystem services are the foundational bases of ecological urbanism that ensure ecological sustainability of Yoruba cities. While MEA defines ecosystem services as "the benefits people obtain from ecosystems" (MEA 2005, p. 1) its 'fragmented' approach can be considered as reductionist. It appears to draw heavily from reductionism where "individual components of wider socio-ecological system are primarily studied in isolation" (Ainscough et al. 2018, p. 94). According to Costanza et al. (1997, p. 253) ecosystem services including goods are "the benefits human populations derive, directly or indirectly, from ecosystem from ecosystem functions". They are "the conditions and processes through which natural ecosystems, and the species that make them up, sustain and fulfill human life" (Daily 1997, p.3). McDonald et al. (2005) define green infrastructure from the perspective of their ecosystem services as the interlocked system of

> waterways, wetlands, woodlands, wildlife habitats, and other natural areas; greenways, parks, and other conservation lands; working farms, ranches and forests; and wilderness and other open spaces that support native species, maintain natural processes, sustain air and water resources and contribute to the health and quality of life (p. 12).

Therefore, reductionism as an ontological approach may not necessarily lead to methodological individualism and methodological individualism can be anti-reductionist and beneficial in social enquiries (Bulle 2019). It is not simply an aggregation of individual components without considering the interaction processes among these constituent parts that give rise to the whole and therefore against the micro-reductionist approach (DeLanda 2019). A fuller explanation of this paradigm and how it influences epistemological and methodological praxis in Yoruba urbanism would be given later in this chapter.

Reductionist ontology implies that ecosystem services of green infrastructure in Yoruba cities should be studied and understood by dividing them into parts that can be singled out for effective study. However, it does not work in such simplistic manner in real life situation since a single ecosystem of green infrastructure can serve multitude of purposes. In the Yoruba cities, a garden tree that produces fruits for human consumption can have its different vegetative parts useful for many purposes. In the process of fruiting, the pollination mechanism may involve certain insects that not only create their own colony in this ecosystem but also feed on the flower nectars. Honey bees could build their combs in the process which are also a rich source of natural honey for human consumption and ailment treatments. The bark of the tree trunk and its leaves can serve as medicinal component for treating other types of human ailments. While all these are still on provisioning, a supporting role has been involved in the case of the honey bees, and generally supporting the ecosystem to perform the three other roles of provisioning, regulating, and cultural services. The same tree also provides regulating service while supporting the human and nonhuman species. It controls the balance of carbon dioxide and oxygen in the air for human survival as it releases oxygen in its metabolism process. Simultaneously, as a regulating service, the tap and fibrous roots of the same tree act as binders to the soil and thereby inhibit soil erosion by surface runoff water. The vegetative foliage of the tree serves as wind breaker against turbulent winds that accompanies thunderstorm rainfalls in this tropical climatic zone. In terms of cultural services which are nonmaterial, the same tree has aesthetic and recreational benefits and could support scientific learning about the natural world. In the Yoruba cities, especially at the urban cores, trees could serve traditional spiritual purposes that could be metaphysically fetish.

In view of these reasons, a conceptual framework that is based on a combination of reductionism and the second ontological paradigm, holism, theories would be suitable for assessing the ecosystem services of green infrastructure in Yoruba cities. Perrotti (2020) argues that coalescing reductionism and holism aligns with the vision of Odum (1977) on the new ecology to bridge the gap between social and natural sciences as applicable in ecological urbanism now. Holism is an ontology for knowledge generation that allows a comprehensive analysis of multifaceted systems like ecosystems. According to Ainscough et al. (2018) it is a "complex systems approach, aiming to understand environmental, social, economic and political aspects of a situation (and the interactions between these)". It is amenable to "a plurality of different knowledge types [...] considered from diverse academic disciplines and local, indigenous and traditional knowledge" (p. 94). The Indigenous Knowledge System embodied in this paradigm makes it suitable as a front leg in a holism-reductionism mixed ontological praxis as a lens for exploring ecological urbanism of Yoruba cities. This considers its grass-root bottom-up approach. Holism theory is part of the complexity science for examining complex systems in nature where it signifies that portions of a whole are in close interconnection, such that they cannot exist autonomously of the whole (del Cerro Santamaría 2020).

The idea of the city as a precarious space for governmentality of urban ecosystems, conservation, and nature are well rooted in political ecology (Vaccaro et al. 2013). In this realm, how eco-justice is achieved in the city is crucial. It deals with deep emphasis on how nature should be incorporated, configured, manicured, adapted, and accepted in the city to achieve eco-city, ecological sustainability and human wellbeing (Pow 2018). These would have bearings on ecosystem services and biodiversity policies (Leibenath 2017). For the city to be truly ecological, adequate emphasis on an urbanism that allows eco-justice systems where human and nonhuman ecosystems are accorded their rightful places in governmental policies is crucial. The city also draws strength from the contrasts of both dynamism and permanence to achieve spatio-ecological balance. While there is a level of permanence of the urban built fabric, at the same time a high status of dynamism characterizes the natural ecosystem of the city. Floral and fauna lives vary in population, location, species, and other ecosystem profiles with some threatened by extinction thereby affecting their ecosystem services. Yoruba cities not only match these descriptions, they are sufficient for studying the parameters from a distinct socio-cultural perspective.

In view of ontological conceptualizations of Yoruba cities, the urban spaces assessed in the study reported in this book qualify as cities. These cities have ecosystem services of green infrastructure that are situated within urban contexts. Therefore, towards developing a conceptual framework, there is the need to provide brief insights into the Yoruba cities whose ecological urbanism is narrated. This is to juxtapose these conceptualizations on the urbanism of the cities and their fittingness for discourses on ecological urbanism. Aside from these realities, it is important to recall the words of William Russell Bascom (1912–1981) on qualifying Yoruba urban places as cities. Bascom (1955) emphasizes as far back as around mid-twentieth century that:

> The Yoruba of Western Nigeria have large, dense, permanent settlements, based upon farming rather than upon industrialization the pattern of which is traditional rather than an outgrowth of acculturation. They are undoubtedly the most urban of all African peoples, the percentage living in large communities being comparable to that in European nations (p. 446).

If this assertion was logically accurate then, the Yoruba cities are of no less significance in this twenty-first century and more importantly two of the cities, Lagos and Ibadan, are currently of megacity status.

1.3.2 *Epistemologies of Ecosystem Services of Green Infrastructure in Yoruba Cities*

The goal in this section is to attempt answering the second question on the epistemologies of studying ecosystem services of green infrastructure in Yoruba cities. Epistemology is a philosophical understanding of the relationship between the truth (existence or reality) and observer. Epistemology emanates directly from ontology of truth, an arm of science that has to do with the nature of knowledge that could be derived from the truth. It has to do essentially with the source, configuration, possibility, and boundary of valid human knowledge. Epistemology therefore helps the legitimization of knowledge about truth. It has been defined by various scholars in urban studies and other areas of knowledge in different ways. For instance, Harding and Blokland (2014) conceptualize epistemology as "a view about the nature of knowledge; about how we know what we know and what counts as evidence in confirming or refuting claims to knowledge" (p.13). Killam (2013) argues that "epistemology examines the relationship between knowledge and the researcher during discovery"(p.8). Epistemology explains how knowledge is created by making clear distinctions between object and subject. In these terms, object is referred to as the material or immaterial substance on which the truth is desired. Subject refers to the observer or searcher of the truth about the object. Through this approach, epistemology is categorized into three groups that are also connected to their root ontologies. These are objectivism, constructivism, and subjectivism. Objectivism posits that truth resides exclusively inside the object irrespective of the subject. Constructivism recognizes Truth as 'meaning'. It postulates that meaning (truth) about an object is 'constructed' by the subject through the process of interaction between the object and subject. Subjectivism advances that truth (meaning) exists inside the subject who imposes the truth on the object. In these three scenarios, objectivism insists there is one and only one truth about any object and therefore rooted in realist ontology. Accordingly, both constructivism and subjectivism maintain that there are multiple truths about any object depending on the subject and therefore rooted in relativist ontology (Moon and Blackman 2014).

There is the need to explain the position of the participants in the research of this book to develop and legitimize a conceptual framework. Their position is based on the Indigenous Knowledge System (IKS) on ecosystem services of green infrastructure in Yoruba cities. Ecosystem service is a component of environment-behaviour research. The study of environment-behaviour is a social research agenda that is not amenable to the rigidity of singular truth embedded in objectivism. Since environment-behaviour research is essentially about finding the outcome of the interplay between objects and subjects respectively, it is well suited for constructivism. The extremity of subjectivism in insisting that truth resides exclusively inside the subject and not as the outcome of interplay between object and subject makes it unsuitable for the research presented in the book. Therefore, in view of these reasons and the highlighted ontological position in this study, the default constructivist epistemology is suitable for developing a conceptual framework. This is because the research is based on the need to understand the ecological relationship between Yoruba people and the green infrastructure in Yoruba cities. This epistemology also accounts for the author's reflexivity as a non-participant observer in the study. While this book is outside the confines of the 'epistemological wars' on conceptual-methodological discourses of realism-relativism (Raskin 2001), it is imperative to affirm that it is guided by

> the basic conviction that when we turn to the examination of those concepts that philosophers have taken to be the most fundamental – whether it is the concept of rationality, truth, reality, right, the good, or norms – we are forced to recognise that in the final analysis all such concepts must be understood as relative to a specific conceptual scheme, theoretical framework, paradigm, form of life, society, or culture (Bernstein 1983, p. 8).

The foregoing suggests that the social relativism-constructivism (Raskin 2001) account presented in this book is sufficient for understanding socio-cultural concepts of Yoruba cities and their ecological urbanism. This epistemological position for studying Yoruba ecological urbanism is preferred to the realism-objectivism paradigm. The perspective aligns with Raskin (ibid) who argues that:

> scientists do not come to know the world through anything but their constructions of it, and that one's constructions are not only guides to future actions, but also means of justifying and remaking understandings of past behaviour (p. 291).

In view of these ontological backgrounds, it is convenient to proceed to the epistemologies of green infrastructure and their ecosystem services in Yoruba cities.

From the perspective of constructivist epistemology, green infrastructure in Yoruba cities is laden with many relative theories which can be concurrently true. In a similar manner, constructivist urban ecosystem services are understandable from pluralistic ecosystem theories. In constructivist epistemology, green infrastructure in Yoruba cities can be conceptualized through biodiversity theories (neutral theory) and landscape ecology theory. Hierarchy theory (Wu 2008) is one of the most important landscape ecology theories where urban green infrastructure are in nested spatial hierarchies of grass covers, shrubs, and trees. These can occur in clusters to form landscapes. Such clusters make gardens, groves, and parks in Yoruba cities. There are also tree-lined streets as products of recent urban renewal projects. Tropical rainforest trees that are singly existing or have been conserved for cultural, traditional religion, or recreation purposes in the core neighbourhoods of some of the Yoruba cities are in existence. Another relevant theory is the social-ecological systems theory that posits that "social systems are linked to their specific ambient ecological system in such a way that both systems together form a co-evolutionary, self-organising unit called landscape", a total human ecosystem (Kirchhoff et al. 2013, p. 37). The social system of the Yoruba people is highly compatible with the green infrastructure that co-evolved to form a social-ecological system that is highly referred in time and space (Fadamiro and Adedeji 2016). Landscape heterogeneity theory posits that parts of a landscape differs from one another and a landscape is composed of different elements that make it varied and complex (Zhao et al. 2002; Gastauer et al. 2021). This is closely linked with patterns theory where spatial and temporal heterogeneities lead to different landscape patterns (Gao et al. 2018). The patterns of green infrastructure in Yoruba cities are not only varied in terms of location, morphology, and characterization, they are synonymous to the sociocultural living patterns of the Yoruba people which are ecological (Adedeji and Fadamiro 2018).

Constructivist approach to understanding urban ecosystem services is apparently multifarious simply from the multiple perspectives generally embedded in the widely accepted Millennium Ecosystem Assessment (MEA 2005). Within this framework, the social-ecological system is central to the constructivism of ecological urbanism. The social-ecological system is distinguished as "an integrated system of people and nature in which feedbacks occur between human and biophysical system elements" (Cumming 2014, p.4). The plurality of social lifestyles and urban nature present a constructivist human-environment milieu. Different members of the society derive different ecosystem services from different green infrastructure with numerous viewpoints, perceptions, and realities in Yoruba cities.

1.3.3 Methods of Assessing Ecosystem Services of Green Infrastructure in Yoruba Cities

In view of the ontological and epistemological considerations of assessing ecosystem services of green infrastructure in Yoruba cities already discussed, the aim in this section is to explain how these philosophies guide specific decisions on the research methods adopted. In the first instance, these decisions are based on theoretical perspective of deduction. Through deduction, the earlier highlighted theories, especially the established Millennium Ecosystem Assessment (MEA 2005), are tested on the Yoruba cities that constitute the study population. This is unlike elsewhere where induction is not only novel but the only appropriate research orientation. Mike Allen (2018, p.694) posits that "induction refers to the practice of inferring general claims from regularities observed within a particular body of data [*and*] such general claims are usually delivered as probabilistic judgements or law-like statements" for obtaining empirical knowledge. Inductivism, a mode of logic reasoning based on induction, is essentially an objectivist method of enquiry. It relies on probability sampling with the goal of generalizing results over the study population based on quantitative statistical techniques. Induction would not be suitable for this research in view of its realism ontology, objectivism epistemology, and positivism (including post-positivism and structuralism) philosophies. Induction is for the purpose of prediction to arrive at generalization. It was not used for this research design of assessing ecosystem services of green infrastructure in Yoruba cities.

The reasons are not farfetched. The induction categories are generally broad strokes that are rigidly based on statistical logics, assumptions, and laws and providing sketchy monolithic insights without in-depth consideration of unique contexts. Induction leads to quantitative research designs. Quantitative techniques involve definition of the research problem, devising research questions, setting up aim and objectives, and delineating the research population. Others are the determination of sampling frame, sample size, sampling techniques, operationalization of variables, methods of data collection, and statistical data analyses (Adedeji 2019). However, the socio-ecological focus of this work is rooted in the intricacies of human behavioural lifestyles in their ecological milieu which are not amenable to quantitative techniques of inductive enquiries. The weakness of inductive reasoning lies in its inability to guarantee its conclusions, which are probabilistic for scientific inferences and decision-making processes (Hayes and Heit 2018). It only supports its conclusions based on the contexts of the debates and how they are operationally delineated which makes it unsuitable for assessing ecological urbanism of Yoruba cities.

On the other hand, deduction is most useful for making valid inferences that are not based on any probability but on set of premises. It has one-hundred percent assurance and guarantee being specific about object and subject simultaneously (Hayes and Heit 2018). Deduction is a logical reasoning that posits that valid knowledge can be generated by focusing on specific and applicable laws existing in human, nonhuman, and social processes to test their truthfulness over a population. It is based on interpretivism and the proponents are called interpretivists whose pursuit in research is comprehension. This is by observing individual circumstances to trace the evolution of phenomena qualitatively (Crotty 1998). The specificity of focus in terms of object and subject from problem definition to inference through data collection and interpretation makes deduction a non-probabilistic 'scientific' approach whose result could be guaranteed. The interpretivist' mechanism of deduction makes it unsuitable for the research methods in the natural sciences. Conversely, the rigid research methods in natural sciences through inductive contextualization of positivism (and post-positivism, and structuralism) are not suitable for socio-ecological science where "interpretations of reality are culturally derived [*and*] historically situated". This goal of deduction to understand through constructivism (social), interpretivism, hermeneutics, phenomenology, and symbolic-interactionism (Moon and Blackman 2014, p.1169) makes it suitable for developing a conceptual framework for assessing ecosystem services of green infrastructure in Yoruba cities. The goal is explored with the already existing Millennium Ecosystem Services (MEA 2005) of the United Nations. It is also in harmony with the relevant sections and subsections of the United Nations' Sustainable Development Goals of 2030 Agenda (UN 2019). These frameworks provide suitable conceptual starting points for understanding the ecological urbanism of Yoruba cities. The research therefore utilizes *a priori* logics based on theoretical deduction rather than empirical observations in induction.

A third philosophical category called abduction (Osman et al. 2018) combines both induction and deduction elements. This implies that abduction transverses between generalization through empirical sampling (induction, empirically-driven research approach) and testing such generalized theories again (deduction, conceptually-driven research approach) in a continuous cycle until a 'grounded' stage is reached. This accounts for the fittingness of abductivism in grounded theory research designs. Since the aim and objectives of studying ecosystem services of green infrastructure in Yoruba cities are achievable through deduction, abduction was not considered necessary for developing a conceptual framework in the research design.

Even though knowledge generation in the present age is at the intersection of multi- and inter-disciplinary engagements, the phenomenon of ecological urbanism of Yoruba cities is essentially anthropological. This implies the cultural and biological study of human societies (both rural and urban) in the past and present including their norms and values in their respective environments (Marcus and Fischer 2014). Since ethnography is a suitable methodology in anthropological enquiries (Mitchell 2017), it not only reinforces the plausibility of this research design, it convincingly situates the whole endeavour in the right positions. These include identifying, 'collecting', analysing and interpreting the data through ecological and socio-cultural lenses in the accurate realm. That is, relativist ontology, constructivist epistemology, interpretivist theoretical perspective and methods of studying Yoruba cities and their ecological urbanism. This is because humans are diverse even within the same homogenous ethno-linguistic group like the Yoruba people who vary in dialects of the same Yoruba language. This methodological position is also supported by 'diversity', a component of socio-culturalism in human relationship with the biosphere (Essono 2019). Specifically, metaphysical rationalisation are required in some instances as "explanations of phenomena by any nonmaterial force, such as mind or spirit or a deity- things that, by definition, cannot be investigated by the methods of science" (Bernard 2018, p. 5–6). These become necessary because indigenous Yoruba people sometimes operate with mystical insights in deriving ecosystem services from green infrastructure in Yoruba cities. This research design accounts for the author's reflexivity as a non-participant observer in the narrated study. Therefore, the social relativism/constructivism (Raskin 2001) account presented in this book is endemically acceptable in all socio-ecological discourses on urban dwellers (Mac Callum et al. 2019).

Furthermore, the study of ecological urbanism is readily amenable to qualitative research design that is interpretive since urban places are also eco-culturally construed (Meier and Frers 2016). Hessse-Biber (2018, p. 547) argues that "interpretive approach assumes a subjective reality that consists of stories or meanings grounded in natural settings". Accordingly, this qualitatively-driven research is a situated activity in Yoruba cities of Nigeria and "consists of a set of interpretive, material practice that make the world" of the ecosystem services that Yoruba people derive from urban green infrastructure visible. It is a "naturalistic approach" turning the Yoruba cities into "a series of representations, including field notes, interviews, conversations, photographs and recordings attempting to make sense of, or interpret, phenomena in terms of the meanings people bring to them" (Denzin and Lincoln 2011, p. 3).

This research is about socio-ecological lifestyles embedded in cultural practices of the Yoruba people. The highlighted approaches are in support of the idea that culture is situated in the psyche of social actors (Escobar 2018). Culture is "whatever it is one has to know or believe in order to operate in a manner acceptable to its members in all requisite social situations" (Geertz 2008, p. 43). Because of this engagement with the social domain, the research is without doubt situated in the field of environmental humanities where "a culture can be described by reconstructing categories, taxonomies and systematic rules to produce something akin to ethnographic algorithm" (Gobo 2018, p. 71). It enables the seeking "to understand how people construct the world around them, what they are doing, how they are doing it or what is happening to them in terms that are meaningful and that offer rich insights" into how socio-ecological processes are constituted (Flick 2018, p. 5).

The research instruments were formulated as structured interview guide and ethnographic observation schedule that are based on the concept of ecological urbanism and its green infrastructure in selected Yoruba cities. They were designed based on the parameters of socio-cultural ecosystem services of green infrastructure in the literature (Scott et al. 2016; Kambo et al. 2019; Mell and Clement 2020; Mowatt 2018). The research participants are in four categories of key informants. They are: (a) traditional worshipers at the urban cultural green spaces, (b) officials and government workers managing the green spaces, (c) tourists, recreationists and general users of the green infrastructure, and (d) owners of urban home gardens. Sampling of the interviewees was carried out through simple random method.

Consents for interviews were secured through the established courtesy norms in the Yoruba culture and explanations on the academic intent of the research. Interviews and ethnographic observations were carried out in daylights. The interviews were conducted in both English and Yoruba languages as necessary. Translation is an acceptable procedure in interviews and ethnography, the influence of these on the interview situation being accounted for and minimized (Littig and Pöchhacker 2014). Pseudonyms are used for the participants in the study for anonymity (Goodwin et al. 2020; Chauvette et al. 2019). However, the need to authenticate their narratives on ecosystem services as true-life situations requires the mentioning of their real and specific locations, including brief information on their socioeconomic profiles (Walford 2018; Brear 2018). This is also to show all-inclusiveness in their selections to the extent that is necessary to juxtapose these on their narratives to enhance constructivist' argumentation.

Furthermore, customary status prefixes are added to the names as a sign of respect to the interviewees as part of sociocultural ethics. It is not cultural in the region to refer to unfamiliar people simply by their first names especially if the second person is older in age. The "verbatim transcribed interviews" texts data obtained were analyzed deductively through qualitative content analyses by a process of condensation, coding, categorization, and themeing as recommended by Erlingsson and Brysiewicz (2017, p. 94). Deductive approach offers straightforwardly applied but logical lay down *modus operandi* for analyzing qualitative data that could generate dependable and convincing results (Bernard 2017). These set of procedures are generally called deductive analysis by reading raw data with the focus of identifying emerging themes and concepts by way of interpretation (Merriam and Grenier 2019; Harding 2018). Results were themed qualitatively by identifying spatial and landscape design plant elements that are associated with specific cosmological, gastronomic, and medicinal values. Similar qualitative method was used by Lau et al. (2014) to formulate "framework for healthy campus open space design" (p. 465) from an ethnographic qualitative study and also in many qualitative "conceptual frameworks and models of health and natural environments" in Lawrence and Forbat (2019, p. 148).

The dissertation so far has shed light on the theoretical-methodological approach engaged in this book. The task in the next section of this chapter is to synthesize the key ontological, epistemological, and methodological arguments on ecosystem services of green infrastructure into a conceptual framework for analysing the case studies of Yoruba cities.

1.3.4 Conceptualizing Ecosystem Services of Green Infrastructure in Yoruba Cities

Green infrastructure and ecosystem services have been linked conceptually in the literature (Hansen and Pauleit 2014). For the current study reported in this book, the following emerging conceptual framework emerges from the broad literature considered in the previous sections of this chapter. It describes six main concepts and how they are interlinked in the study presented: ecological urbanism, ecosystems, green infrastructure, ecosystem services, indigenous knowledge system, and Yoruba cities. Lindley et al. (2018) put forward that frameworks are "structured sets of concepts, assumptions and relationships" that could be "conceptual, analytical or process based […] as the foundation for understanding a particular view of reality". They submit further that frameworks are "informed by disciplinary specialisms which reflect the particular worldviews of researchers and their epistemological, ideological or contextual underpinnings" (p. 9). For these reasons, the emerging conceptual framework that guides the study shall consist of the listed six concepts, their components of uneven contours, characterization, and contexts. They inevitably form the research design contextualization in the conceptual framework which is acceptably complex (Jabareen 2009).

The structure of the conceptual framework is shown in Fig. 1.4. It consists of four strata that are gleaned from the literature discussed in the previous sections. These are theoretical underpinnings linking green infrastructure and ecosystem services (Eisenman 2013), philosophical considerations for studying ecosystem services (Boons 2013), research design contextualization (Cilliers 2019; Seiwert and Rößler 2020), and ecosystem components of United Nation 2030 Agenda for Sustainable Development Goals (Maes et al. 2019). The four strata present the indicators and metrics that allow for a bottom-up participatory process for delivery strategies to achieve evidence-based eco-cultural Yoruba cities through a combined feedback-feedforward mechanism. The underlying theories include holism theory (Ainscough et al. 2018), reductionism theory (Perrotti 2020), socio-ecological systems theory (Gonçalves et al. 2021), Millennium Ecosystem Service Assessment, image ability, and Landscape pattern and Heterogeneity theories. A mixed holism-reductionism approach in the framework is adopted by conceptualizing Yoruba cities as single homogenous ethno-linguistic group that has cosmological-ecological commonalities, but spatio-distinctly located geographically. Reductionism is explored in assessing the ecosystem services of green infrastructure in each distinct entities called cities and their individual socio-ecological settings, including individual human actors. Through socio-ecological systems theory, the Yoruba cities are understood as ecosystems of human and nonhuman entities that simultaneously derive and contribute ecosystem services by means of cultural eco-aesthetics to achieve socio-ecological resilient cities (Haruna et al. 2018). These services are characterized according to the Milenium Ecosystem Service Assessment and include provisioning, supporting, regulating, and cultural services. The resultant image ability of the Yoruba cities takes the morphology, identity, and the meanings that indigenous Yoruba people derive from green and blue spaces into account. The image ability leads to place attachment that fosters ecological relationship between humans and nonhumans in the Yoruba cities (Cocks and Shackleton 2020a, b). The identification of distinctiveness of elements of landscapes like trees, shrubs, ground covers, and water and how they are interwoven to form characterizing patterns in landscape heterogeneity theory and landscape pattern theory (Gustafson 2019) make them suitable as part of the theoretical building block for the framework.

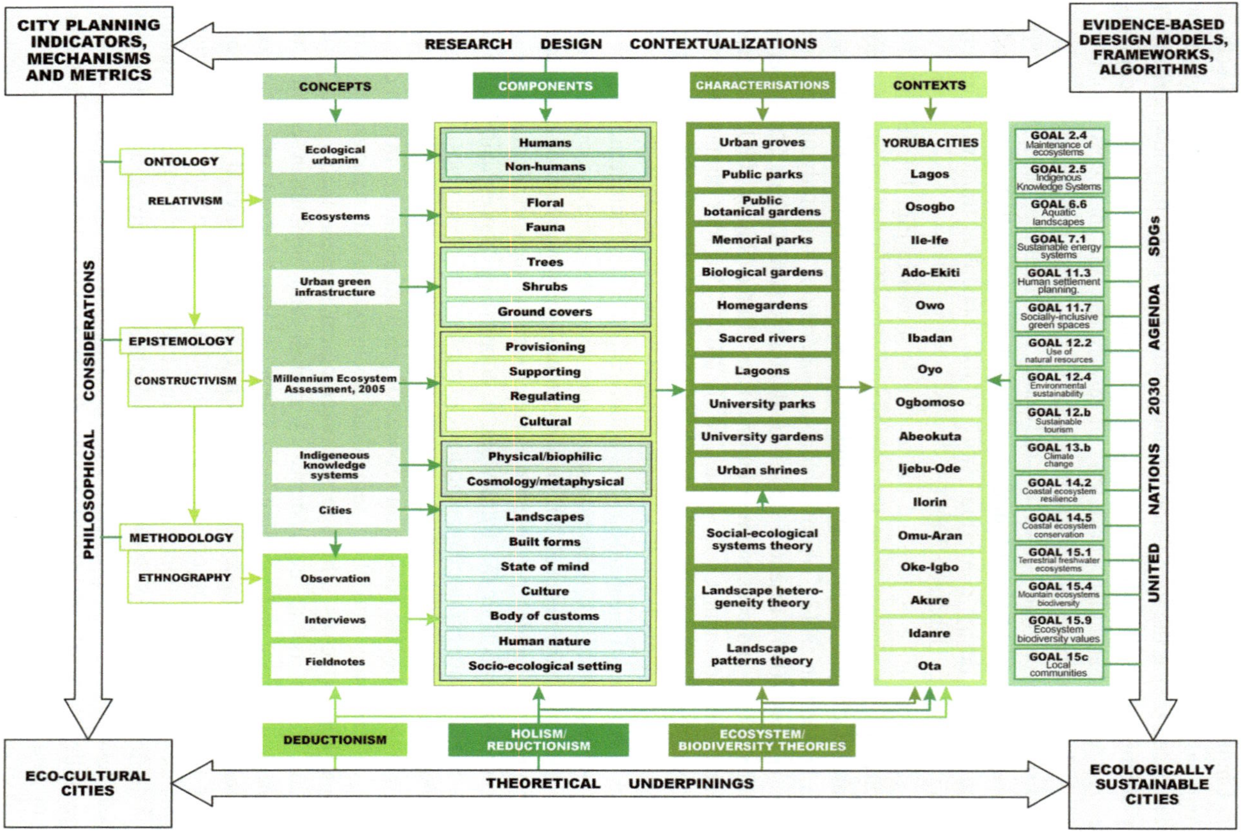

Fig. 1.4 Conceptual framework for analyzing the ecosystem services of green infrastructure in Yoruba cities. Source: Author

The philosophical considerations in the framework are on relativist ontology (relativism), constructivist epistemology (constructivism), and ethnographic methodology (ethnography) (Farrow et al. 2020). These are explored from the deductive (deductionism) perspective of reasoning in the research design contextualisations (Mac Kinnon 2022). The relativist ontology in the framework defines the multiplicity of the components of the concepts of green infrastructure, ecosystems, millennium ecosystem assessment, indigenous knowledge system, and ecological urbanism of Yoruba cities. Green infrastructure is an ecosystem or a network containing a number of ecosystems that have parts and serve multifarious functions to enhance the wellbeing of the system (Li et al. 2020). It is a category of landscapes that have complex strata to enhance ecological association of Yoruba people and their nonhuman environments. They consist of components like trees, shrubs, ground covers, sacred and no sacred water bodies, air, and soil that are characterized as groves, public parks and gardens, home gardens, and all landscape networks in Yoruba cities that are physical or metaphysical. The services, benefits, and functions that the green infrastructure performs are the ecosystem services in the framework. The components of the ecosystem services are provisioning, supporting, regulating, and cultural services (MEA 2005). Provisioning services are the material resources derived by the indigenous Yoruba people from the urban ecosystems for survival. Supporting services comprise the benefits derived from green infrastructure that ensure sustainability of the ecosystems. Regulating services guarantee ecological balance of the ecosystems of Yoruba cities. Cultural services are non-material benefits that are socio-ecological and sociocultural. The interwoven combination of human and nonhuman entitities and their functioning are the evidences of an ecological urbanism of the Yoruba cities that provide the contexts in the framework. The cities include Ibadan, Ado-Ekiti, Ogbomoso, Owo, Akure, Ile-Ife, Osogbo, Lagos, Oyo, Ijebu-Ode, Abeokuta, Oke-Igbo, among others.

The cities are not only geographically diverse, they exhibit the material and nonmaterial qualifications of cities in human civilizations. They have cosmological intricacies of the heavenly celestial and the earthly terestial symbolisms and traditional religious value-system. They are socio-culturally framed as biological entities with biocultural diversities and complex socio-ecological adaptive systems. The scale of the case studies in each city is based on ecological characterizations of green infrastructure. These include urban groves, public parks, public botanical gardens, memorial parks, biological gardens, home gardens, sacred rivers, lagoons, university parks and gardens, and urban shrines. This is in agreement with the socio-ecological systems theory, landscape heterogeneity theory and landscape pattern theory. The characterizations also concur with the classifications of urban open and green spaces by Mumford (1969) and Francis (1991) based on historical origins, Woolley (2003) based on benefits, and Stanley et al. (2012) based on range of forms and ecosystem functions. The characterizations based on functions (Gehl 2011), sub-categories (Shakibi 2015), and spatial design elements (Lau et al. 2014) are also relevant and agree with the characterizations in the proposed framework.

The epistemological constructivism of Yoruba cities in the framework attests to the close-knit diverse ecological relationships between the individual Yoruba people and the components of their urban environments. It is an integrated system of varying lifestyles of Yoruba people and biophysical elements of nature in their environments that are diversely constructed for varying benefits contained in the Millennium Ecosystem Assessment (MEA 2005). The constructivism in this human-nonhuman relationship is evident in the Indigenous Knowledge System (IKS) of the Yoruba people. The IKS is essentially eco-cultural and a foundational component of biological and ecological diversities (Kosoe et al. 2020). It shows how local knowledge is explored to achieve human wellbeing by the Yoruba people through exploration of ecological relationships that leverage on ecosystem services of green infrastructure in the cities (Du Toit et al. 2018). The diversities of these relationships are explored through ethnographic methodology that allows for deep focus on individual human and nonhuman members of Yoruba cities. The methodology in the research design links the concepts that form ecological urbanism (ecosystems, millennium ecosystem assessment, green infrastructure, IKS), their components, and characterization of the components in specific urban contexts (Maciejewski et al. 2021). This is through ethnography which allows for in-depth understanding of the IKS of Yoruba people on how wellbeing goals are obtained through ecosystem services of green infrastructure.

The research design contextualization in the conceptual framework aligns with many of the UN 2030 Agenda for SDGs discussed in this Chapter. Components of the goals are targeted at fostering, enhancing, and supporting the urban ecosystems (Maes et al. 2019). These include goals 2.4, 2.5, and6.6 with details shown in Fig. 1.4. Others are goals 7.1, 11.3, 11.7, 12.2, 12.4, 12.b, 13.b, 14.2, 14.5, 15.1, 15.4, 15.9, and 15c (Yang et al. 2020; Maes et al. 2019; Wood et al. 2018) in the conceptual framework and as earlier discussed in Sect. 1.1 and shown in Fig. 1.4. The developed conceptual framework takes care of the urban ecological realities in Yoruba cities and fittingly situated within global literature. The framework guides responses to the following questions in this book: What are the peculiar characteristics or way of life of the Yorubas that promotes ecological integrity and development? What are their local wisdom, practices, ideologies, customs, tradition, norms, habits, etc., that other people elsewhere can inculcate to promote human ecology? These are evidence-based ecological factors expressed through IKS. The framework also promises to be adaptable to studying ecosystem services of urban green infrastructure in non-Western context because of its biocultural diversity approach. The next chapter of the book discusses the ecosystem services of specific green and blue spaces in particular Yoruba cities, to understand the mechanisms of their performance through the lenses of this conceptual framework in the current chapter.

References

Adedeji JA (2019) Survey research designs incorporating data dummying: the what and how of a strategy for studying environment security. In: Figiel K (ed) SAGE research methods cases part 2. SAGE Publications Ltd, London, pp 1–15

Adedeji JA (2020) Urban biocultural identity of Yorubas: intersection of philosophy and nature for wellbeing at Osun sacred grove UNESCO site, Osogbo, Nigeria. In: Cocks M, Shackleton C (eds) Urban nature: enriching belonging, wellbeing and bioculture, 1st edn. Routledge, London, pp 51–66

Adedeji JA, Fadamiro JA (2018) Urbanisation forces on the landscapes and the changing value-systems of Osun sacred grove UNESCO site, Osogbo, Nigeria. Landsc Res 43(6):798–816

Ainscough J, Wilson M, Kenter JO (2018) Ecosystem services as a post-normal field of science. Ecosyst Serv 31:93–101

Allen M (2018) Induction. In SAGE encyclopedia of communication research methods. SAGE, London

Aluko BT, Omisore EO, Amidu AR (2008) Valuation of Yoruba sacred shrines, monuments, and groves for compensation. In Indigenous peoples and real estate valuation. Springer, Boston, pp 145–173

Amati M (2021) The City and the super-organism. RMIT University, Melbourne

Anderson M (2006) Yoruba: history and culture. Retrieved from hurri.kean.edu/anderson/ppt/worldgeo/Yoruba.ppt

Atanda JA (1996) The Yoruba people: their origin, culture and civilisation. In: Olatunji OO (ed) The Yoruba history, culture and language. Ibadan University Press, Ibadan, pp 3–34

Bairoch P (1988) Cities and economic development: from the dawn of history to the present. University of Chicago Press, Chicago

Barbier EB, Burgess JC (2017) The sustainable development goals and the systems approach to sustainability. Economics 11(1)

Bascom W (1955) Urbanization among the Yoruba. Am J Sociol 60(5):446–454

Bascom W (1962) Some aspects of Yoruba urbanism. Am Anthropol 64(4):699–709

Bayulken B, Huisingh D, Fisher PM (2021) How are nature based solutions helping in the greening of cities in the context of crises such as climate change and pandemics? A comprehensive review. J Clean Prod 288:125569

Benedict MA, McMahon ET (2012) Green infrastructure: linking landscapes and communities. Island press, Washington, DC

Bernard HR (2018) Research methods in anthropology: qualitative and quantitative approaches, 6th edn. Altamira Press, Lanham

Bernard HR (2017) Research methods in anthropology: qualitative and quantitative approaches. Rowman & Littlefield, Lanham

Bernstein RJ (1983) Beyond objectivism and relativism: science, hermeneutics, and praxis. University of Pennsylvania Press, Philadelphia

Bevir M (2021) London: a city of humanism and power. Critical review of international social and political philosophy, 1–20

Bianconi F, Clemente M, Filippucci M, Salvati L (2018) Regenerating urban spaces: a brief commentary on green infrastructures for landscape conservation. TEMA J Land Use Mobil Environ 11(1):107–118

Boons F (2013) Organizing within dynamic ecosystems: conceptualizing socio-ecological mechanisms. Organ Environ 26(3):281–297

Brear M (2018) Swazi co-researcher participants' dynamic preferences and motivations for, representation with real names and (English-language) pseudonyms–an ethnography. Qual Res 18(6):722–740

Bulle N (2019) Methodological individualism as anti-reductionism. J Class Sociol 19(2):161–184

Bullock C, Joyce D, Collier M (2018) An exploration of the relationships between cultural ecosystem services, socio-cultural values and Well-being. Ecosyst Serv 31:142–152

Chauvette A, Schick-Makaroff K, Molzahn AE (2019) Open data in qualitative research. Int J Qual Methods 18:1–6. 1609406918823863

Cilliers EJ (2019) Reflecting on green infrastructure and spatial planning in Africa: the complexities, perceptions, and way forward. Sustain For 11(2):455

Clark T (2020) Ecological grief and anthropocene horror. American Imago 77(1):61–80

Cocks M, Shackleton C (2020a) Situating biocultural relations in city and townscapes: conclusion and recommendations. In: Cocks M, Shackleton C (eds) Urban nature: enriching belonging, wellbeing and bioculture, 1st edn. Routledge, London, pp 241–268
Cocks M, Shackleton C (eds) (2020b) Urban nature: enriching belonging, wellbeing and bioculture, 1st edn. Routledge, London, pp 51–66
Costanza R, d'Arge R, de Groot R, Farber S, Grasso M, Hannon B, Naeem S, Limburg K, Paruelo J, O'Neill RV, others (1997) The value of the world's ecosystem services and natural capital. Nature 387:253–260
Cowgill GL (2004) Origins and development of urbanism: archaeological approaches. Annu Rev Anthropol 33:525–549
Crotty M (1998) The foundations of social research. Sage, London
Cumming GS (2014) Theoretical frameworks for the analysis of social–ecological systems. In: Social-ecological systems in transition (pp. 3–24). Springer, Tokyo
Daily G (1997) Nature's services: societal dependence on natural ecosystems. Island Press, Washington DC
Danermark B, Ekström M, Karlsson JC (2019) Explaining society: critical realism in the social sciences. Routledge, Abingdon
Davies C, MacFarlane R, McGloin C, Roe M (2006) Green infrastructure planning guide. Project: Final Report
del Cerro Santamaría G (2020) Possibilities and challenges for intercultural research in global urbanism. In: Examining cultural perspectives in a globalized world, IGI global, pp 94–115
DeLanda M (2019) A new philosophy of society: assemblage theory and social complexity. Bloomsbury Publishing, London
Denzin NK, Lincoln YS (2011) Introduction: the discipline and practice of qualitative research. In: Norman KD, Yvonna (eds) The Sage handbook of qualitative research, 4th edn. Thousand Oaks, pp 1–20
Dickinson DC, Hobbs RJ (2017) Cultural ecosystem services: characteristics, challenges and lessons for urban green space research. Ecosyst Serv 25:179–194
Du Toit MJ, Cilliers SS, Dallimer M, Goddard M, Guenat S, Cornelius SF (2018) Urban green infrastructure and ecosystem services in sub-Saharan Africa. Landsc Urban Plan 180:249–261
Eisenman TS (2013) Frederick law Olmsted, green infrastructure, and the evolving city. J Plan Hist 12(4):287–311
Erixon Aalto H, Marcus L, Torsvall J (2018) Towards a social-ecological urbanism: co-producing knowledge through Design in the Albano Resilient Campus Project in Stockholm. Sustain For 10(3):717
Erlingsson C, Brysiewicz P (2017) A hands-on guide to doing content analysis. Afr J Emerg Med 7(3):93–99
Escobar A (1999) After nature: steps to an antiessentialist political ecology. Curr Anthropol 40(1):1–30
Escobar A (2018) Culture, economics, and politics in Latin American social movements theory and research. In: The making of social movements in Latin America, Routledge, pp 62–86
Essono A E (2019) Link to the biosphere: man, condemned to alterity and coviability. In: Coviability of social and ecological systems: reconnecting mankind to the biosphere in an era of global change, Springer, Cham, pp 583–609
Fadamiro JA, Adedeji JA (2012) Cultural morphology of cityscapes: a semiotic analysis of Osogbo, Nigeria. LAMBERT Academic Publishing, Saarbrücken
Fadamiro JA, Adedeji JA (2016) Cultural landscapes of the Yoruba of South-Western Nigeria demystified as solidified time in space. Space Cult 19(1):15–30
Falola T, Genova A (2005) Orìsà: Yoruba gods and spiritual identity in Africa and the diaspora. Africa World Press, Trenton
Fandrich IJ (2007) Yorùbá influences on Haitian vodou and New Orleans voodoo. J Black Stud 37(5):775–791
Farrow R, Iniesto F, Weller M, Pitt R (2020) GO-GN research methods handbook

Ferreira V, Barreira AP, Loures L, Antunes D, Panagopoulos T (2020) Stakeholders' engagement on nature-based solutions: a systematic literature review. Sustain For 12(2):640
Flick U (2018) Doing qualitative data collection – charting the routes. In: The SAGE handbook of qualitative data collection. Sage, London, pp 3–16
Francis M (1991) Urban open spaces. In: Zube EH, Moore GT (eds) Advances in environment, behaviour, and design, vol 1. Plenum Press, New York\London
Gao H, Sabo JL, Chen X, Liu Z, Yang Z, Ren Z, Liu M (2018) Landscape heterogeneity and hydrological processes: a review of landscape-based hydrological models. Landsc Ecol 33(9):1461–1480
Gastauer M, Mitre SK, Carvalho CS, Trevelin LC, Sarmento PS, Meira Neto JA et al (2021) Landscape heterogeneity and habitat amount drive plant diversity in Amazonian canga ecosystems. Landsc Ecol 36(2):393–406
Gavrilidis AA, Popa AM, Onose DA, Gradinaru SR (2022) Planning small for winning big: small urban green space distribution patterns in an expanding city. Urban For Urban Green 78:127787
Gates C (2011) Ancient cities: the archaeology of urban life in the ancient near east and Egypt, Greece and Rome. Taylor & Francis, London
Geddes P, LeGates R, Stout F (2021) Cities in evolution. Routledge, London
Geertz C (2008) Thick description: toward an interpretive theory of culture. In: The cultural geography reader, Routledge, pp 41–51
Gehl J (2011) Life between buildings: using public space. Island Press, Washington DC
Gill SE, Handley JF, Ennos AR, Pauleit S (2007) Adapting cities for climate change: the role of the green infrastructure. Built Environ 33(1):115–133
Girard LF (2016) The city and the territory system: towards the "new humanism" paradigm. Agric Agric Sci Proc 8:542–551
Gobo G (2018) Upside down – reinventing research design 1. In: the SAGE handbook of qualitative data collection, SAGE Publications Ltd., London, p 65–83
Golley FB (1991) The ecosystem concept: a search for order. Ecol Res 6(2):129–138
Gonçalves P, Vierikko K, Elands B, Haase D, Luz AC, Santos-Reis M (2021) Biocultural diversity in an urban context: an indicator-based decision support tool to guide the planning and management of green infrastructure. Environ Sustain Indic 11:100131
Goodwin D, Mays N, Pope C (2020) Ethical issues in qualitative research. Qual Res Health Care:27–41
Gordon JU (1979) Yoruba cosmology and culture in Brazil: a study of African survivals in the new world. J Black Stud 9(4):231–244
Gustafson EJ (2019) How has the state-of-the-art for quantification of landscape pattern advanced in the twenty-first century? Landsc Ecol 34(9):2065–2072
Hagan S (2014) Ecological urbanism: the nature of the city. Routledge, Abingdon
Hansen R, Pauleit S (2014) From multifunctionality to multiple ecosystem services? A conceptual framework for multifunctionality in green infrastructure planning for urban areas. Ambio 43(4):516–529
Harding A, Blokland T (2014) Urban theory: a critical introduction to power, cities and urbanism in the 21st century. Sage, London
Harding J (2018) Qualitative data analysis: from start to finish. Sage, London
Haruna AI, Oppong RA, Marful AB (2018) Exploring eco-aesthetics for urban green infrastructure development and building resilient cities: a theoretical overview. Cogent Soc Sci 4(1):1478492
Hayes BK, Heit E (2018) Inductive reasoning 2.0. Wiley Interdiscip Rev Cogn Sci 9(3):e1459
Hessse-Biber S (2018) Toward an understanding of a qualitatively driven mixed method data collection and analysis: moving toward a theoretically centered mixed methods praxis. In: The SAGE handbook of qualitative data collection, Sage. pp 545–563
Hoddy ET (2019) Critical realism in empirical research: employing techniques from grounded theory methodology. Int J Soc Res Methodol 22(1):111–124
Howard E (1946) Garden cities of tomorrow. Faber, London, p 3
Hull RW (1976) Urban design and architecture in precolonial Africa. J Urban Hist 2(4):387–414

Jabareen Y (2009) Building a conceptual framework: philosophy, definitions, and procedure. Int J Qual Methods:49–62

Jacobs J (2016) The death and life of great American cities. Vintage, New York

Jones S (2002) Social constructionism and the environment: through the quagmire. Glob Environ Chang 12(4):247–251

Jörg K (2019) Nietzsche and ecological reason (s) in the Anthropocene. Trumpeter: J Ecosophy 35(1):22–40

Kadykalo AN, López-Rodriguez MD, Ainscough J, Droste N, Ryu H, Ávila-Flores G et al (2020) Disentangling 'ecosystem services' and 'nature's contributions to people'. Ecosyst People 15(1):269–287

Kambo A, Drogemuller R, Yarlagadda PK (2019) Assessing biophilic design elements forecosystem service attributes – A sub-tropical Australian case. Ecosyst Serv 39(100):977

Killam L (2013) Research terminology simplified: Paradigms, axiology, ontology, epistemology and methodology. Laura Killam, Sudbury

Kim J, Son Y (2021) Assessing and mapping cultural ecosystem services of an urban forest based on narratives from blog posts. Ecol Indic 129(107):983

Kirchhoff T, Trepl L, Vicenzotti V (2013) What is landscape ecology? An analysis and evaluation of six different conceptions. Landsc Res 38(1):33–51

Kosoe EA, Adjei POW, Diawuo F (2020) From sacrilege to sustainability: the role of indigenous knowledge systems in biodiversity conservation in the Upper West Region of Ghana. Geogr J 85(4):1057–1074

Kwaja C (2011) Nigeria's pernicious drivers of ethno-religious conflict. Africa Centerfor Strategic Studies, Washington, DC

Landscape Europe News letter (2015, January) Retrieved 19 August, 2015 from http://www.landscape-europe.net/images/LEN_January_2015.pdf

Lau SSY, Gou Z, Liu Y (2014) Healthy campus by open space design: approaches and guidelines. Front Arch Res 3(4):452–467

Lawrence RJ, Forbat J (2019) Rethinking conceptual frameworks and models of health and natural environments. Health 23(2):158–179

Le Corbusier L (1980) Urbanisme. Arthaud

Lefebvre H, Kofman E, Lebas E (1996) Writings on cities, vol 63. Blackwell, Oxford

Leibenath M (2017) Ecosystem services and neoliberal governmentality–German style. Land Use Policy 64:307–316

Leong DP (2012) Street signs: toward a missional theology of urban cultural engagement, vol 12. Wipf and Stock Publishers

Li T, Dong Y, Liu Z (2020) A review of social-ecological system resilience: mechanism, assessment and management. Sci Total Environ 723(138):113

Lindley S, Pauleit S, Yeshitela K, Cilliers S, Shackleton C (2018) Rethinking urban green infrastructure and ecosystem services from the perspective of sub-Saharan African cities. Landsc Urban Plan. https://doi.org/10.1016/j.landurbplan.2018.08.016

Littig B, Pöchhacker F (2014) Socio-translational collaboration in qualitative inquiry: the case of expert interviews. Qual Inq 20(9):1085–1095

Lynch K (1960) The image of the city. MIT Press, Cambridge, MA

Mac Callum D, Babb C, Curtis C (2019) Doing research in urban and regional planning: lessons in practical methods. Routledge, New York

Maciejewski K, Currie P, O'Farrell P (2021) Social-ecological connectivity in Global South cities. In: Urban Ecology in the Global South. Springer, Cham, pp 347–364

Mac Kinnon NJ (2022) Reductionism: friend or Foe of an Integrative Social Psychology? Soc Psychol Q 85(1):6–22

Maes MJ, Jones KE, Toledano MB, Milligan B (2019) Mapping synergies and trade-offs between urban ecosystems and the sustainable development goals. Environ Sci Pol 93:181–188

Maffi L (2007) Biocultural diversity and sustainability. In: Pretty J, Ball AS, Benton T, Guivant JS, Lee DR, Orr D, Pfeffer MJ, Ward H (eds) Sage handbook on environment and society. SAGE, London, pp 267–277

Mao Q, Wang L, Guo Q, Li Y, Liu M, Xu G (2020) Evaluating cultural ecosystem services of urban residential green spaces from the perspective of residents' satisfaction with green space. Front Public Health 8:226
Marcus GE, Fischer MM (2014) Anthropology as cultural critique: an experimental moment in the human sciences. University of Chicago Press
Marinic G (2012) Ecological Urbanism. Des Issues 28(1):105–106
McDonald L, Allen W, Benedict M, O'connor K (2005) Green infrastructure plan evaluation frameworks. J Conserv Plan 1(1):12–43
McHarg IL (1969) Design with nature. American Museum of Natural History, New York, pp 7–17
Meier L, Frers L (2016) Encountering urban places – Visual and material performances in the city. In: Encouraging urban places, Routledge, pp 17–24)
Mell I, Clement S (2020) Progressing green infrastructure planning: understanding its scalar, temporal, geo-spatial and disciplinary evolution. Impact Assess Proj Apprais 38(6):449–463
Mell IC (2010) Green infrastructure planning–integrating connectivity and multifunctionality with linear and large scale developments. Landsc Archit China 9(1):131–143
Mell IC (2013) Can you tell a green field from a cold steel rail? Examining the "green" of Green Infrastructure development. Local Environ 18(2):152–166
Mell IC (2017) Green infrastructure: reflections on past, present and future praxis. Landsc Res 42(2):135–145
Merriam, S. B., & Grenier, R. S. (Eds.). (2019). Qualitative research in practice: examples for discussion and analysis. John Wiley & Sons 2.
Millennium Ecosystem Assessment, MEA (2005) Ecosyetem and Well-being: synthesis. Island Press, Washington, DC
Mitchell JC (2017) On quantification in social anthropology 1. In: The craft of social anthropology. Routledge, pp 17–46
Moon K, Blackman D (2014) A guide to understanding social science research for natural scientists. Conserv Biol 28(5):1167–1177
Moore MJ (2009) Pluralism, relativism, and liberalism. Polit Res Q 62(2):244–256
Morris WF, Ehrlén J, Dahlgren JP, Loomis AK, Louthan AM (2020) Biotic and anthropogenic forces rival climatic/abiotic factors in determining global plant population growth and fitness. Proc Natl Acad Sci 117(2):1107–1112
Mostafavi M (2010) Why ecological urbanism? Why now. Eco Urbanism:12–53
Mostafavi M, Doherty G (2010) Ecological urbanism. Lars Müller, Frankfurt
Mostafavi M, Doherty G (2016) Ecological Urbanism in Latin America. urbe. Revista Brasileira de Gestão Urbana 8:7–11
Mowatt RA (2018) Understanding Ifá: inserting knowledge of an African cosmology in leisure studies and nature-based research. Leis Stud 37(5):515–532
Mumford L (1937) What is a city. Archit Rec 82(5):59–62
Mumford L (1938) The culture of cities. Harcourt, Brace and Co., New York
Mumford L (1961) The city in history: its origins, its transformations, and its prospects, vol 67. Houghton Mifflin Harcourt
Mumford L (1968) The urban prospect. Harcourt Brace Jovanovich, New York
Mumford L (1969) The philosophy of urban open space. In: Seymour WN (ed) Small urban spaces: the philosophy, design, sociology, and politics of vest-pocket parks and other small urban open spaces. New York University Press, New York, pp 13–22
Mumford L (2016) The culture of cities, vol 19. Open Road Media
Neumann B, Ott K, Kenchington R (2017) Strong sustainability in coastal areas: a conceptual interpretation of SDG 14. Sustain Sci 12(6):1019–1035
Nilsson A, Vanhuyse F, Arra V, Requena A, Agerström M (2021) Citizen participation for sustainable cities. Stokholm Environment Institute. Accessed from https://cdn.sei.org/wp-content/uploads/2020/03/factsheet-citizen-participation-for-sustainable-cities.pdf on 12 June 2022.
Obateru OI (2006) The Yoruba city in history: eleventh century to the present. Penthouse, Ibadan
Odum EP (1977) The emergence of ecology as a new integrative discipline: ecology must combine holism with reductionism if applications are to benefit society. Science 195(4284):1289–1293

Olmsted FL (1870) Public parks and the enlargement of towns. American Social Science Association, Cambridge, MA, pp 52–99
Opoku A (2019) Biodiversity and the built environment: implications for the Sustainable Development Goals (SDGs). Resour Conserv Recycl 141:1–7
Osman S, Mohammad S, Abu MS, Mokhtar M, Ahmad J, Ismail N, Jambari H (2018) Inductive, deductive and abductive approaches in generating new ideas: a modified Grounded Theory study. Adv Sci Lett 24(4):2378–2381
Padoa-Schioppa C (2017) La menteecologica del Landscape Urbanism. Ri-Vista Res Landsc Archit 15(2):230–239
Park RE, Burgess EW (2019) The city. University of Chicago Press, Chicago
Perrotti D (2020) Urban metabolism: Old challenges, new frontiers, and the research agenda ahead. In: Urban Ecology. Elsevier, pp 17–32
Perrotti D, Iuorio O (2019) Green infrastructure in the space of flows: an urban metabolism approach to bridge environmental performance and user's wellbeing. In: Planning Cities with Nature. Springer, Cham, pp 265–277
Pickett ST, Kolasa J, Jones CG (2010) Ecological understanding: the nature of theory and the theory of nature. Elsevier
Porpora DV (2018) Critical realism as relational sociology. In: The Palgrave Handbook of Relational Sociology. Palgrave Macmillan, Cham, pp 413–429
Potschin M, Kretsch C, Haines-Young R, Furman E, Berry P, Baró F (2016) Nature-based solutions. In: Potschin M, Jax K (eds) Open NESS ecosystem services reference book. EC FP7 Grant Agreement no. 308428. Available via: www.openness-project.eu/library/reference-book
Pow CP (2018) Building a harmonious society through greening: ecological civilization and aesthetic governmentality in China. Ann Am Assoc Geogr 108(3):864–883
Raskin JD (2001) On relativism in constructivist psychology. J Constr Psychol 14(4):285–313
Register R (1987) Ecocity Berkeley: building cities for a healthy future. North Atlantic Books, Berkeley
Reimer JM, Devillers R, Claudet J (2021) Benefits and gaps in area-based management tools for the ocean Sustainable Development Goal. Nat Sustain 4(4):349–357
Riechers M, Barkmann J, Tscharntke T (2018) Diverging perceptions by social groups on cultural ecosystem services provided by urban green. Landsc Urban Plan 175:161–168
Rising HH (2019) The image of the water city. J Urban Des 24(3):424–442
Säumel I, Reddy SE, Wachtel T (2019) Edible City solutions – One step further to foster social resilience through enhanced socio-cultural ecosystem services in cities. Sustain For 11(4):972
Schröter M, Van der Zanden EH, van Oudenhoven AP, Remme RP, Serna-Chavez HM, De Groot RS, Opdam P (2014) Ecosystem services as a contested concept: a synthesis of critique and counter-arguments. Conserv Lett 7(6):514–523
Scott M, Lennon M, Haase D, Kazmierczak A, Clabby G, Beatley T (2016) Nature-based solutions for the contemporary city/Re-naturing the city/Reflections on urban landscapes, ecosystems services and nature-based solutions in cities/Multifunctional green infrastructure and climate change adaptation: brownfield greening as an adaptation strategy for vulnerable communities?/ Delivering green infrastructure through planning: insights from practice in Fingal, Ireland/ Planning for biophilic cities: from theory to practice. Plan Theory Pract 17(2):267–300
Seiwert A, Rößler S (2020) Understanding the term green infrastructure: origins, rationales, semantic content and purposes as well as its relevance for application in spatial planning. Land Use Policy 97(104):785
Shakibi N (2015) A proposed framework for analyzing the spatial organization of urban open space networks: case studies: Tehran and Isfahan, Iran (Doctoral dissertation, University of British Columbia).
Smith ME (2005) City size in late postclassic Mesoamerica. J Urban Hist 31(4):403–434
Spirn AW (1984) Granite garden. Basic Books
Spirn AW (2005) Ecological urbanism: a framework for the design of resilient cities. Landsc Res 30:359–377

Spirn AW (2014) Ecological urbanism: a framework for the design of resilient cities (2014). In: The ecological design and planning reader. Island Press, Washington, DC, pp 557–571
Stanley BW, Strak BL, Johnston KL, Smith ME (2012) Urban open spaces in historical perspective: a transdisciplinary typology and analysis. Urban Geogr 249, 33(8):1089–1117
Steiner F (2011) Landscape ecological urbanism: origins and trajectories. Landsc Urban Plan 100(4):333–337
Tadi M, Manesh, S. V. (2014) Transformation of an urban complex system into a more sustainable form via integrated modification methodology (IMM). Int J Sustain Dev Plan 9(4):514–537
Tansley AG (1935) The use and abuse of vegetational concepts and terms. Ecology 16:284–307
Till KE (2001) New urbanism and nature: green marketing and the neo traditional community. Urban Geogr 22(3):220–248
UN (2019) Sustainable development goals. Retrieved from https://www.undp.org/sustainable-development-goals
Vaccaro I, Beltran O, Paquet P (2013) Political ecology and conservation policies: some theoretical genealogies. J Polit Ecol 20(1):255–272
Waldheim C (2006) The landscape urbanism reader. Princeton Architectural Press, New York
Walford G (2018) The impossibility of anonymity in ethnographic research. Qual Res 18(5):516–525
Wallace KJ (2007) Classification of ecosystem services: problems and solutions. Biol Conserv 139(3–4):235–246
Ward C (1978) The child in the city. Architectural Press, London, p 73
Wilk B, Säumel I, Rizzi D (2021) Collaborative governance arrangements for co-creation of NBS. In Nature-based solutions for more sustainable cities–A framework approach for planning and evaluation. Emerald Publishing Limited
Wirth L (1938) Urbanism as a Way of Life. Am J Sociol 44(1):1–24
Wood SL, Jones SK, Johnson JA, Brauman KA, Chaplin-Kramer R, Fremier A et al (2018) Distilling the role of ecosystem services in the Sustainable Development Goals. Ecosyst Serv 29:70–82
Woolley H (2003) Urban open spaces. Spon Press, London
World Population Review, WPR (2021). Accessed from https://worldpopulationreview.com/countries/nigeria-population
Wright FL (1971) The language of an organic architecture. Structurist 11:80
Wu J (2008) Toward a landscape ecology of cities: beyond buildings, trees, and urban forests. In: Ecology, planning, and management of urban forests, Springer, New York, NY, pp 10–28
Xiao L, Haiping T, Haoguang L (2017) A theoretical framework for researching cultural ecosystem service flows in urban agglomerations. Ecosyst Serv 28:95–104
Yang S, Zhao W, Liu Y, Cherubini F, Fu B, Pereira P (2020) Prioritizing sustainable development goals and linking them to ecosystem services: a global expert's knowledge evaluation. Geogr Sustain 1(4):321–330
Zhao Y, Yu X, Guang W (2002) Review on landscape heterogeneity. Ying Yong Sheng tai xue bao = The. J Appl Ecol 13(4):495–500

Chapter 2
Green-Blue Spaces in Yoruba Cities – Ecosystem Services Ethnography

Abstract Ecological urbanism combines the principles of ecology and city design to pursue urban environmental quality for human wellbeing. It sits at the intersection of ecosystem services (ES) of urban green and blue spaces (UGBS). Ecological urbanism combines all material and metaphysical elements of nature for the wellbeing of city dwellers. This chapter presents results of the assessment of ES of specific UGBS in selected Yoruba cities using the socio-cultural ecological conceptual framework developed in Chap. 1. The conceptual framework links ecological urbanism, ecosystems, green infrastructure, ecosystem services, indigenous knowledge system, and Yoruba cities together. The assessment takes care of theoretical underpinnings linking green infrastructure and ecosystem services, philosophical considerations for studying ecosystem services, contextualization of the concepts, and ecosystem components of United Nations 2030 Agenda for Sustainable Development Goals. The assessment results are presented as metanarratives of the ES of specific UGBS in selected Yoruba cities. Yoruba cities are precolonial urban spaces with unique ES of UGBS. The aim of this chapter is to underscore the intricate ecological connections of these spaces to the everyday life of the citizens and their understandings of urban natural processes through indigenous knowledge system and local wisdom. This is in view of the constantly emerging and evolving urban biogeophysical crisis. To achieve this, Osun Grove UNESCO World Heritage Site in Osogbo was selected as archetype that represents the divinatory cosmology and pre-colonial city planning in Yoruba land. Its ES reveal a belief-system in the river and medicinal plants that is rooted in a mind-body connection with the spirit world of the goddess and Yoruba pantheon in general. The Grove demonstrates the traditional spirituality and religiosity of Yoruba people. Typical of the Grove, other green spaces like Biological Garden and Park in Akure, Lekki Conservation Centre in Lagos, Fajuyi Park Ado-Ekiti, and three University parks and gardens, are sites of provisioning, supporting, regulating, and cultural ES. The urban ecotourism values of some of the sites contribute to human wellbeing aside from being part of nature-based solution to the environmental and economic sustainabilities of the cities. As sites of social capital, memorials, metabolic and aesthetic appreciations of nature, they nurture symbiotic relationships between human and nonhuman natures that are reminiscent of Garden of Eden. The chapter concludes with a section on the Lagos lagoon and its

J. A. Adedeji, *Ecological Urbanism of Yoruba Cities in Nigeria*, Cities and Nature, https://doi.org/10.1007/978-3-031-34688-0_2

associated University of Lagos Park as a unique urban ecosystem. It offers recommendations on the strategies for sustaining the ES of the ecological urban spaces. The next chapter of the book discusses the diverse ways Yoruba people relate with Green Environment to access their ES for human wellbeing and the mechanisms that frame these services.

Keywords Socio-ecological ecosystem services · Osun Osogbo Grove UNESCO Site · Urban parks and gardens · Human wellbeing · Garden of Eden · Experience of nature

2.1 Cultural Green Spaces as Urban Ecotourism Destinations in Yoruba Cities

Urban Green-Blue Spaces (UGBS) are green infrastructure and water bodies of various categories that showcase the ecological urbanism of Yoruba cities. They are public UGBS that occur in three categories. These are cultural green spaces; English parks, gardens, and memorials; and Institutional parks and gardens. Water bodies could be located in these spaces or not. This chapter contains metanarratives of the ecosystem services of these three categories and water bodies, drawn from different selected cities in the region. The cultural green spaces in Yoruba urbanism have unique ecosystem services that are eco-cultural. In addition to being socio-cultural spaces and generally cultural landscapes offering socio-ecological ecosystem services, they offer provisioning, regulating, and supporting ecosystem services. The goal here is to identify these services generally through a meta-analysis of interview responses.

2.1.1 Osun Grove UNESCO World Heritage Site, Osogbo, Nigeria

Osun Grove is the most popular urban grove in Nigeria. It is a naturally endowed urban ecological site that was inscribed on the UNESCO World Heritage list in 2005 as archetypal of Yoruba divinatory cosmology and pre-colonial city planning. The Grove contains 42 traditional worship shrines, flora of primary rainforest, and River Osun. The city of Osogbo was birthed in the grove. In view of its enlisting, the provisioning ecosystem services of the grove have been limited due to the legal framework for its ecological conservation. Mrs. Alice Ijaoye, a Bachelor of Science degree holder, at the age of 45 years was the Chief Education Officer and key informant at the Grove. In addition to the inscription of the rules guiding the conduct of visitors at the site, she emphasized some of the rules during an interview. According to her, "The rule is that you must not cut any tree, you must not farm, you must not kill any animal here and you must not kill the fishes in the river. Dumping of refuse is not allowed and building of house". She revealed that the grove is home to about 400 plant species out of which about 250 are medicinal, the River Osun water being believed to be natural medicinal concoction. "That is what Osogbo people use to bathe their babies for healing", she submitted. She explained further that the people of Osogbo do not believe in treating their children in health clinics but rather with herbal bathing and administering concoction for wellness. This cultural practice has been inscribed in their everyday life to the extent of translating it into a socio-cultural vernacular song that says "*seleru agbo, agbara agbo, l'osun fi n wo omo re ki dokita o to de*". It simply means that Osun goddess is implied to be a mother that nurtures her children with bathings of herbal concoctions before the advent of western healthcare experts. To affirm this point at the moment of the interview, she pointed at a traditionalist preparing herbal concoction at the Grove – "so they are preparing the herb in order to drink. I think you are seeing this woman preparing it? Referring to Osun as the "goddess of fertility", she noted the belief of the traditional adherents that the goddess "is the giver of children, for any barren woman that comes here, it depends on their faith. So faith is everything in life". This belief-system is rooted in a mind-body connection with the spirit world of the goddess and Yoruba pantheon in general. The core adherents of Christianity and Islam in Osogbo do not subscribe to this belief-system, except those who are practicing multiple religions (Adedeji 2020).

Mrs. Alice Ijaoye described the ecosystem services of the Grove in a broad sense to showcase its multifarious ecological significance:

> People normally come for adventure, sightseeing and relaxation. When they see it, [...] it makes them happy. Number two is that people normally come for spiritual aspects of it. They normally come to consult the Osun [*goddess*], they normally come to take their spiritual bath. Yes, and all the trees and leaves, they are all medicinal. [...]People normally come, people that know leaves, how to use them, they normally come and source them and use it for whatever they want to use it for. And the water itself is concoction. They don't call it water when you get there, they call it concoction. They call it *Agbo*. If you come here and tell them you want to come and fetch water here, they will ask you to go back and that there is no water here. But if you call it concoction, they will ask you to go and fetch the concoction, for those that believe in it, it works for them.

The custodians of the Grove are always present at different worship points and at the access point to the Osun River. They are traditional adherents of the Osun cult system who have oversight of the Grove despite the administrative presence of UNESCO and International Council on Monuments and Sites (ICOMOS). The Grove is also an urban eco-tourism site. Through this ecosystem service, it draws multitude of local, national, and international tourists to Osogbo on a daily basis with the peak period during the Osun Osogbo annual festival. Mr. Yemi Adeyeri is a Yoruba man from the Adagba Royal Compound in Ile-Ife. He was over 40 years of age at the time of interviewing him in the Grove as a key informant, being a Museum Education Officer and Head of the Department. He passionately discussed the socio-economic ecosystem services of the Grove as a bye-product of its urban ecotourism value. Mr. Yemi Adeyeri proceeded as follows:

> Let me start from economic benefit. It's one of the values that we derive here. Just now I said you will pay something, and the money you pay is to whose account? Federal Government [*of Nigeria*] account. No matter how small. Yoruba will say drops of water make the river. I collect ten Naira from you, and ten [*Naira*] from there, if 'am able to collect ten Naira from one million people [*tourists*], that means I've obtained 10 million [*Naira*]. So also, is it here. Whatever you pay now, some other people will pay their own. So economic benefits to the government is in various ways: 1) from the proceeds of the revenue; 2) from the tax collected from the hotels and other places. When visitors [*tourists*] come, they would have to lodge in hotels if they cannot go back to where they came from. They cannot sleep under the tree. And definitely, when the hotel is being patronized, to pay tax to the government would not be so difficult.

His economic analyses of revenue from this site is characteristic of the multiple layers of urban ecotourism and their concomitant benefits. The benefits have ripple effects on the members of the host communities and urban governance. Jamali et al. (2021) identified the socioeconomic overflows of urban ecotourism to include promotion of economic advancement for individuals in host communities, collective communal economic prosperity, and boosting of morale. Others are raising capitals for investment, empowerment of local communities, and promotion of respects for cultures of host communities. Since economy is one of the three pillars of sustainable tourism, the other two being social and environmental (Machnik 2021), Osun Grove is well suited for a typical desirable urban ecotourism site, being nature-based, and an ecological unit of Osogbo city. Urbanecotourism provides practical opportunities for human and nonhuman nature to interact, understand natural processes, and internalize the essentialities of biophilic relationships. Machnik (2021) argues that ecotourism is embedded with appreciation for the aesthetics of nature and reverence for the traditions and culture of indigenous societies. Accordingly, Steiner (2014) links ecotourism, urban ecology, and ecological urbanism together within a landscape ecological urbanism paradigm. Within this paradigm, urban economy offers a sustainable *conditio sine qua non* to which urban ecotourism comes to the rescue. This is by leveraging on numerous economic outcomes in many city sectors, including transportation, local food marketing, and merchandize in everyday utility items. To this end, Mr. Yemi Adeyeri emphasized further that:

> [...] you boarded a vehicle from your destination to this place, either from Ilesa, from Akure or from anywhere. You have increased the sales or the profit of that transporter. You didn't come to Osogbo to come and greet me, neither did you come to the market, you're in Osogbo because of the Osogbo Sacred Grove. So, you can see that it attracts so many things. What about the food seller, the local food vendor there, *Mama put*, when you're hungry now, you have to go and buy. If she has targeted twenty customers today, [...] it will increase to twenty-eight. You're not in Osogbo because of her, and not because her food is very good. Definitely her sale will increase. So also for everybody that lives in this environment.

He explained further that:

> Whenever you [*they*] see visitors [*tourists*] coming in you see them bringing banana, orange, people that were not selling pure [*potable sachet*] water before, they turn to pure [*potable*] water seller and those that were not selling coke [*soft drink*] before turn to coke seller all because they know that people are coming down here and they will surely drink something. So, in terms of economic benefit, I cannot explain everything. I cannot tell you

> this is how much. Even in times of festival, you see people selling kegs [*plastic containers*], remember we have *agbo*[concoction]here, the river is concoction and people want to fetchfrom it. So, you see keg seller, *kampala* [traditional fabric] seller and craftsmen.

The foregoing details the many layers of huge economic enhancements that the socioeconomic ecosystem services of Osun Grove impact on Osogbo, Osun State and Nigeria at large as a global urban ecotourism destination. It has and continues to boost the hospitality industry of the city. The quality and quantity of hotel services have significantly improved with attendant sophistication of the urban systems. There is a vast informal trade comprising all kinds of locally made souvenier items dominated by artistic items, local fabric industry, and social products like traditional music and musical instruments. In his narration, Mr. Yemi Adeyeri noted the centripetal power of the Grove that "because of this grove alone, we are able to accommodate so many people that wouldn't have thought of coming to Nigeria." These ecotourism adventures present opportunities for enhancing social capital of the urbanites and cultural exchanges between the local and global communities. Accordingly, Mr. Yemi Adeyeri continued to reveal "And when they come, there are lot of things we learn from each other. There is no way you meet somebody that you will not learn something". Through these windows of opportunities, Osun Grove contributes in no small measure to the quality of the socio-ecological milieu of Osogbo. It affords recreation in a natural environment with serenity and highest obtainable air quality compared with the hustling and bustling of the city bound neighbourhoods, thereby enhancing the health conditions of tourists. For these reasons and from many interviews conducted at the site, the site is always patronized by all categories of tourists of wide-ranging socioeconomic statuses. In addition, Osun Grove is akin to a museum of natural history, cultural laboratory, healing garden, and outdoor laboratory for all aspects of the basic and applied sciences. The diversity of religious inclinations of tourists, Osun custordians, and traditionalists makes the Grove a theological site. Accordingly, Mr. Yemi Adeyeri added that "Even in religious studies, the theologians want to come and compare traditional religion and Christian religion, Islam with Christian religion. They want to see the relationship among all these."

Despite all these multifarious ecosystem services of the Osun Grove, interview respondents reported ecologically-related disservices. One of the commonly reported ecosystem disservices of the Grove is phobia for its nonhuman nature including snakes like python, monkeys, and the River Osun. Others are fear of the shrines, sculptures, and insect bites. This is in view of the primary nature of the forest landscape and its animistic sculptures. According to Mr. Yemi Adeyeri:

> And the fear of the tradition is always there too. You may be seated and they bring rituals in the day, even in the night. But as the tradition, the ritual you're not supposed to see will not be brought in the day light. They carry them in the night. And if you walk in the night, you'll see strange things you're not supposed to see.

Mrs. Alice Ijaoye also emphasized that "The only thing I fear is the insect, we have a lot of them" and "[…] you can just be going and see snakes and you will run back. […] We have some little little flies, insects that will just bite, those are the challenges". Koole and Van den Berg (2005) report the dark side of wild nature of conservation areas, woods, and bush to be feared. This is in association of wilderness with death reminders as people come to grips with some of their deepest existential fears (Solomon et al. 2004). Research (Sonti et al. 2020) has shown that fear of forest as an ecosystem disservice is an important concern that precludes some urban dwellers from visiting natural sites in cities. Reconciling ecosystem services and disservices has been argued to be essential to the formulation of an ecosystem framework that "suggests a renewed focus on whole-of-ecosystem function." This is on the premise that both 'extremes' are laden with simplicity and "ecological ambiguity." The premise rests on the claim that any interaction between the simplistic notions of ecosystem services and ecosystem disservices "can be positive, negative, or neutral (from an anthropocentric perspective)" (Saunders and Luck 2016, p.1363) as a result of ecological trade-offs that configure ecosystem processes. One thing that is evidently clear in the case of Osun Grove is that, in the face of persistent patronage, globalized popularity, and sustained cultural prowess, the identified disservices not only become insignificant but have also been ignored in the human and nonhuman interactions in its ecological urbanism. No human continues to patronize a place without any benefit.

2.2 Biological Garden and Park, Akure, Nigeria

Akure is a medium sized city with a population of 360,268 in 2006 (NPC 2006) and projected to be 690,533in 2021 according to the World Population Review (WPR 2021). Akure has been the capital city of Ondo State since 1976 when the State was created. The ecological setting of the city is characterized by two main stretches of urban forests, the Federal College of Agriculture forest and the Ministry of Agriculture Biological Garden and Park located along Oba Adesida Road. A third park of mini-status is the Fasoranti Park opposite the Governor's Office at Alagbaka. The Biological Garden and Park is situated at a prime location within the urban form and is replete with both indigenous and exotic species of trees, in a combination of wild and cultivated urban forests. These are different from the horticultural section that hosts plant nurseries, adult recreation ground, and playground for children with some amusement facilities. The plants' nursery section is for the horticultural purpose of producing seedlings for sale to the public for multiple purposes including homegardening, ornamental plantings, and general landscaping. While the wild forest has trees like Iroko (*Milicia excelsa*, F. Moraceae) and Obeche (*Triplochiton scleroxylon*, F. Malvaceae), the dominant cultivated forest consists mainly of teak (*Tectona grandis*, F. Lamiaceae) plantation. The site was managed by the Ondo State Ministry of Agriculture as an urban forest conservation but now by the Ondo State Ministry of Culture and Tourism. The Biological Garden and Park is operated as a relaxation centre with numerous urban ecosystem services discussed in this sub-section.

Mrs. Victoria Jekoyemi, a superintending officer of the Garden, a Yoruba, at the age of 55 years who holds a Bachelor of Science degree in Forestry and Postgraduate Diploma in Sericulture was interviewed at the site. Her narratives of the ecosystem services of the Garden were sketched around regulating and cultural services. To her, the site "serves as carbon sink for [*the...*] environment, for reducing air pollution, [...] for educational purposes, [... *and*] for research." She explained further that the site serves for the "protection of wildlife species" and "provision of employment". Mr. Vincent Iyando is a single, male, Ijaw, Yoruba, English graduate from University of Benin. He was above the age of 35 years when interviewed at the Biological Gardens as a tourist. Vincent approached the description of the ecosystem services of the site from a theological perspective. To him, the Garden "gives you a feeling that Adam and Eve had in the Garden of Eden. We were made to understand in the Bible that Adam and Eve stayed in a place that was very natural. Each time I come here I used to have that feeling". Another interview respondent, male, married, Christian, and hydrologist by profession who serves a Non-Governmental Organisation also discussed the significance of the site from the perspective of the Garden of Eden, notions of Adamic moral fall and its ecological consequences. Through a perceptive mental picture, he sees

> the Garden of Eden as a beautiful one with a perfect arrangement. But by virtue of our human nature, we missed the glory and since then, we have been wallowing in abject poverty. God has made a perfect arrangement for us but due to our behaviour, disobedience and nature, it disabled us. That's why we find ourself in this position.

This insight into the perception of splendor of the Biblical Garden of Eden draws on the common religiosity about Divinity in the Yoruba land. Another tourist affirmed the role of this spiritualisation of nature in the environmental ambience of the Biological Garden and Park, Akure. According to him, "Because God is omnipotent, ominiscient, omni-everything, He knew the type of environment He wanted to create, that's why He created it for us". This attribution of the quality of every aspect of the Garden to the characteristic providence of God suggests the spiritual biophilic lifestyles of Yoruba people. In these submissions, a distinction is made between two types of nature based on their locale – human nature and the nonhuman nature – and that the Garden of Eden represents the divinely ordained best ecological milieu for humans. The disobedient human nature here essentially refers to the nature 'within' humans and the nonhuman nature as the nature 'outside' humans. The disobedient inner nature lead to the alienation of humans from the nonhuman nature of the Garden of Eden where all ecological conditions for sustainable human living were present, and eventually driven out. It also portrays the enstrangement of nature from the very idea of cities as built forms where 'outer/outside' nature has been done away with. However, the Biological Garden, Akure, is an incorporation of nature into the city fabric. A Yoruba, female, Bachelor of Science graduate respondent at the age of 24 years noted, "Everywhere is green, and I know that green is nature. God is awesome." Her understanding conflates nature with the green landscape and attributes both to the ingenious power of the divine God. Describing a multispecies urban ecology of the Garden, a Christian female interviewee at the age of 22 years who hails from another location in Ondo State but resides in Akure approached the subject from a religious perspective by citing the authority of God from the Bible. To her, "The Bible says that [*all that*] God created are beautiful and wonderfully made, and everything that God created are [sic] for our own benefit. The plants we see around us are for our health. Even the animals are for our own benefit and are perfect for us". Her insights concurrently portray many aspects of human-nature relationships, the details of which are not intended here. However, Zylstra et al. (2019), p. 1) aver in a recent nuanced response to the emerging questions of "connectedness with nature' (CWN) or 'human-nature connections (HNC)" that recalls CWN as being:

> a stable state of consciousness comprising symbiotic, cognitive, affective, and experiential traits that reflect, through consistent attitudes and behaviours, a sustained awareness of the interrelatedness between one's self and the rest of nature (Zylstra et al. 2014, p. 126).

The authority of God whose divine prerogative binds all human and nonhuman nature together presents many ecological realities beyond human ingenuity and superfluity of knowledge. All ecological designers cite the authority of God as the source of their inspirations (Spirn 2002; Pickett et al. 2004), like McHarg (1997, p. 321) who insists that ecology is "not only an explanation, but also a command". In line with these ecological connections with nature that the Biological Garden stimulates, the interviewee went on to add that the "place is recreative [... *and*] someone can actually come here to read and study because this place is quiet and one can really think" and be free from environmental stressors. It is "a serene area where people come to relax" and most people visit there for "photo shoot and other activities, do get-together and so on". Also, for Mrs. Oyinade Akinkunmi, a married Christian Yoruba woman of less than 40 years of age and resident in Akure, the site offers natural background for photography and is a natural garden imbued with ambience of fresh air. Mr. Ojo Samuel, a married, Christian, Yoruba, and a university degree holder equally appraised the site as being excellent for his profession of photography by adding beauty which made his customers to "request for the garden because it's so beautiful and because they cherish it a lot." Adeboye, a 20 years old lady held a similar view that people can also patronize the garden just to "cool their head". The aesthetics of nature (Parsons 2007; Dow 2022) and appreciation of natural environments have been linked with human cognition, mental wellness, and overall quality of life (Van Dijk-Wesselius et al. 2018). They are basic requirements of humans to function psychologically for life satisfaction and productive lifestyles in socio-cultural endeavours (Wong et al. 2018; Hurly and Walker 2019).

Similarly, a married female tourist who holds a Bachelor of Science degree and above 30 years of age drew relationships between God, nature, trees, and life with some particularities on 'goodness' and perfection. To her, the "place is a garden, people come here to relax, to feel the natural breeze of the environment" because "what God created is very good. All these things are natural. Particularly, trees give life." Also, "you feel the free air of nature" aside from learning about and identifying types of trees like "Iroko tree, mango tree." What a description, what an ecosystem service! Her description echoes some of the key debates in ecocriticism, "a contraction of ecological literary criticism" (Hamon 2018, p. 63), where life is studied in all its expressions, both metaphorically and literally. Accordingly, Rival (2021) argues about the symbolic significance of trees as reflecting the impulse of human nature to express innate notions with the aid of outward and physical ciphers. The idea of Nichols (2011, p. 3) about "romantic 'nature' characterized by dynamic links among all living things" is an exemplar in such impulse. The whole gamut reverts back to the Edenic story where "the tree of life" reflects "divine prerogatives, immortality and wisdom" (Mettinger 2007, p. 5). These concepts have direct bearings on literal ecocritical value about an intrinsic connection and interdependence between humans and nonhumans (Oppermann 2011, p. 25).

Furthermore, according to Mr. Vincent Iyando, another interviewee, the site has educational ecosystem service about nonhuman species of nature that "We see the squirrels live anytime we come here. It's educational. And the species of trees, you don't see them in other places except here. It's as if God specifically kept these species in this place". This tells the story of the conservation ecosystem service of the site. Urban parks have been identified as veritable tools for biodiversity conservation in cities (Santos et al. 2019; Palliwoda et al. 2017; Wang et al. 2021) thereby reconnecting humans to the nonhuman nature for wellbeing. Mr. Vincent Iyando assessed the site as a space for re-living the drinking lifestyle of Yoruba forefathers thereby fostering a reconnection to the cultural value-system in an urban context. According to him,

> The palm wine joint and the use of calabash in drinking it gives that feeling of what our [*fore*]fathers did. It has increased my knowledge because some some of these things we were hearing it, but when you come here you see them and feel them. You know all this palm wine is a natural and cultural drink of our people.

Mr. Tola Adeyeba was also a tourist interviewed in the Biological Garden. His extensive insights into the ecosystem services of the site show that he is not a novice on ecosystem services in ecological urbanism discourse. To him, the site has ecological significance for the local, national and international communities, connecting humans to the natural environment. He argues that the site "serves as a medium of refreshment for the local people here and environment". He amplified this human-environment connection, arguing that: "I believe that it connects one to nature. For instance, now, you can see the breeze. You can see this whole place filled with trees. One can also understand the symbiotic relationship that is meant to exist between nature and human beings". The inalieniable relationships among air, flora, and humans are evident here. Mr. Vincent Iyando expressed similar thoughts on "the freshness, the serene ambience, the tranquility" of the Garden which make life to be at peace compared with the built components of the city. Mr. Alade Ayedogbon, a Yoruba, male, student and a Christian who resides outside Akure in another city of Ondo State was interviewed in the Biological Garden. He expressed his perception that "the plants here, [*they*] bring a cool atmosphere. It makes it more natural. With the plants here, this place is quite cool. It's more interesting to breadth in fresh air unlike when you are in the [*built*] city [*environment*]". He offered a reflective description of the physiological satisfaction of having "cold or chilled drink under a tree". This is "very interesting coupled with the sound of birds under a cool tree. It is something pleasant and very interesting", he affirmed. Again, the narrative portrays conviviality as a significant factor in ecosystem service of urban greens and to what extent arboreal soundscapes can contribute to human aural sensory ecology. Knowlton et al. (2021) likewise advance the contributions of forest habitat-related supporting services for birds and the values of birds in the ecosystem food web. In corollaryMr. Alade Ayedogbonpointed out that the trees of the Biological Garden"host insects and some of these insects are edible like the grasshopper, we eat it in our village. These insects will also multiply and grow with the cover of the trees." These descriptions are reminiscent of supporting and provisioning ecosystem services. For the reasons advanced, he "wish[*ed*] there are more of this [*garden*] in every part of the country. The breeze, the warm feeling of the trees, the sound of birds are so interesting". Evidently, the experience of natural landscape is the best precursor for human wellbeing and sustainable ecological urbanism. Fisher et al. (2021) discover significant relationships among birdsong, naturalness of environments, and human wellbeing as products of urban biodiversity conservation that has value for city planning and sustainable practices.

Similarly, Tunrayo was a female English undergraduate student at Adekunle Ajasin University Akungba-Akoko above the age of 20 years. She was of Igbo extraction but lived in Akure when interviewed at the Biological Garden. Her perspectives and insights into the ecosystem services of the Garden interlace many layers summarizable as a multifarious ecological capital. While noting that even without the presence of any natural water body in the site, it is a fully shaded place that "provides a very cool and serene environment." Theartistic sculptural asset of the site and "the major natural resources [*of*] trees" make it outstandingly "a cool environment." She maintained that the site offers recreational bar services because of its bar, barbeque, drinks, youghurt, and lots more, that generate income. These make the Garden a site of very high social significance that enhances human interaction and for people to relax, meet and familiarize with other people, and get their minds off work and "from the troubles of life" and "just unwind." What a great place to be in a troubled world of worries and anxiety, as she explained – "By the time you're leaving here, you feel refreshed". A similar view was expressed by Teniayo, a single Yoruba, male Christian, and undergraduate student between the age of 30 and 35 years old. He perceives the Garden as a picnic site that isvery suitable for fun because "it's a very cool and serene environment." Mr. Adeogun Ijaola, a Christian Yoruba man at the age of 40 years, equally recognized the supporting and regulating ecosystem services of the Garden as enhancing air quality, aeration, and oxygen content for human metabolic needs. Another tourist similarly explained the benefits – "I have natural air; I don't need a fan, I don't need AC [*air conditioner*]. I just came over here to [...] have a conversation with my friend." Akinsetan, a male, married, Yoruba, and Christian, was resident in Akure at the time he was interviewed in the Garden. To him, the site is an antidote against depression and a natural cooling ecosystem of trees that settles the mind as a relaxation and fun center. Akinsetan specifically visited the site to make business connections with his automobile business partners but also enjoyed its convivial affordances through repeated visits for play and passive recreation. The claim of Akinsetan that the site offers opportunities for business connections in his automobile business proved to be correct as John, who turned out to be a car dealer, was randomly selected for an interview. John emphasizes the recreational ecosystem qualities and services of the site as a fun center that many other tourists have underscored. He pointed out that the site offers economic benefits to all the stakeholders involved in its administration, including the onsite business operators, Akure Local Government, and Ondo State Government. John remarked that the site is for all comers, including children who patronize it for picnics and artists who shoot movies there. He reframed the use of the site as a place for cultural reconnection with the forefathers of Yoruba people who enjoyed maximum recuperative ecosystem services under tree canopies in their farms. The Yoruba forefathers did not use modern Air Conditioners, and John believes that Air Conditioners are part of the environmental risk factors that could trigger respiratory ailments.

The site provides an ecological balance to the built environment of the city. This was the perception of Mr. Ezekiel Adebare, a 25 years old postgraduate student, single, Yoruba and a resident of Akure. In his assessment that was based on repeated visits, the site is a very nice place where tourists free themselves "from the stress of the city" to refresh their "mind and brain" and the plants are "in their natural environment." To him, "even when seeing the colour green, it has some effects on our mental health", helping to improve it, in addition to socio-cultural exchanges with other tourists "from different backgrounds" as an opportunity "to learn new things from them" and vice versa.

Another male interviewee at the site who was a graduate from Kogi State University at the age of 18 years confirmed his experience of tranquility as he said, "everywhere [*is*] serene" and "I've enjoyed the fresh air, the cool breeze". The garden trees function as air purifiers for human respiratory metabolism at a very high quality index compared with the built city neighbourhoods. The ecological environment of the Garden is exemplarily imbued with many affective qualities that aid tourists and users to recuperate. These place qualities that are affective include relaxing-ability, fascinating-ability, enjoy-ableness, restfulness, inviting-ability, and inspiring-ability. Others are beautifulness, exciting-ability, recuperative-ability, therapeutic-ability, restorative-ability, pleasantness, and comfortableness (Adedeji et al. 2020). Mr. Tola Adeyeba gave a vivid illustration of these symbiotic ecological affordances of the Garden as follows:

> For instance, imagine that I work in [*the*] First Bank, which is just down there at the roundabout, in Alagbaka there. So, imagine I have been so stressed from overworking lately and I need a place to rest – not to eat, but to rest. Instead of going to places like Chicken Republic, I can just find my way here to enjoy the cool breeze of this serene environment.

Aside from emphasizing the restorative-affective quality of the site from physical and mental exhaustion, his evaluation of the affordances of the ecosystem services connected these services to urban locational morphology. The expansive site is not only located within the main urban hub of the city and surrounded by government offices, ministries, parastatals, banks, among others, but easily accessible from all directions through the Oba Adesida double-carriage way. The implication is that ecosystem services of public urban green spaces are easily enhanced by spatial factors like accessibility, proximity, walkability, connectedness, continuity, and convenience. Analyses of these were portrayed by Olaposi. Olaposi was in his mid-twenties and holds a Bachelor's degree, a Christian, Yoruba, and a resident of Akure at the time he was interviewed at the site. He affirmed the natural status of the Biological Garden and poured encomium on its therapeutic landscape that "naturally we all know the beauty of nature and what it does to our health. It's really a place you just come and relax and have some sightseeing. Nature has a way of helping our health". He anchored his opinion on the beneficial impact of recreation in green spaces where human nature

> could get the direct oxygen from the direct sources in the natural environment here rather than the one that has been polluted that we get from the [*environments of*] cities. Recreation is a form of exercise for us. It gives us fresh idea to move on in life. It's really a place to come and escape the rigour of the city.

During these recreation activities, social interaction and opportunities for cultural transfusion are offered because "in rare occasion you can meet foreigners [*white people*] because they enjoy places like this and you can enjoy from their wealth of knowledge" including other Nigerians of different sociocultural backgrounds. For Adesogo, a Yoruba woman at the age of about 50 years holding a Master's degree, the site is "very rare" and "one in town" where "the trees give cool air" making it a natural sanctuary for her. She further asserted "when there's any problem outside, I do come here to relax myself". In view of the neighbourliness of the site to most of the government public offices in Akure, she expressed similar observation for urban governance workers like:

> directors, the DGs [*director generals*] and the likes that have a lot of things to do in their work place, when there is any serious problem, they come here to cool. By the time they are going back to their offices, they would have been refreshed to give their best.

She equally narrated how she has sourced herbal materials from the site for her personal ethnobotanical medication for wellbeing. She noted that the site "has healed many from physical and spiritual [*ailments*]", just as "many have been delivered from the hand of sickness" and "many get to know many good things" through engagements with the site. Furthermore, Esther a single, Christian, resident in Oye-Ekiti and holds a Senior Secondary Certificate Examination (SSCE) was at the age of 35 years when interviewed in the Biological Garden. She had been a regular tourist at the site. She described the ecological significance of the site and other green spaces in the region as natural endowment and sources of fresh air for refreshing relaxation. She emphasized that "when people are stressed from work", they recover at the site, aside from sightseeing animals from in the forest parts of the site, and enjoying "relationship with others". According to her, and as evident in the profiles of the other tourists interviewed in the site, "most people that come here are educated people, so seating down with them, they say somethings that enlighten me and build me up." She also illustrated her opinion with cases of some people who had been sick and they had "been on that problem for years, and someone just mentioned the solution to that problem, [*medicinal*] leaves that they could be put together that will heal them". Equally, Tolu Ariyibi, a Yoruba traditionalist at the age of 23 years and resident in Akure was one of the people who operate a business venture at the site when interviewed there. While noting the nickname of the Garden as *Omi Eja* (simply meaning fish water), she attested to its natural tranquility "because there is no noise here, everything is quiet, anything you want to eat you'll find, for instance pepper soup, and *suya*, and the people selling makes ends meet from this place". In addition, the site has offered her connections to many people among whom are the 'who is who' in the society, because, according to her, "the people that come here are educated, they are not mean people, they are big men, it made me [*to*] know great people".

These are products of urban landscape planning that resulted from the political will in urban governance, the site being established during the Ondo State governorship era of late Chief Adebayo Adefarati (14 February 1931 to 29 March 2007) between 1999 and 2003. Fasihi and Parizadi (2020) equally discovered that locational qualities and configurations are not only important for the utilization of urban parks, they underscore spatial equality of urban dwellers. An essential aspect of this spatial configuration is vitality (Zhu et al. 2020) in measuring intangible ecosystem services of urban parks which is closely linked with conviviality (Barker et al. 2019; Rodriguez and Simon 2015). Hinchliffe and Whatmore (2006) provide deep understanding of the intricate relationships among urban nature, urban wildlife, urban greeneries, and urban ecology, in complimentarity with conviviality while narrating stories of living cities.

Vitality and conviviality not only enhance the quality of urban parks and their significance as ecological recreational spaces (Tibesigwa et al. 2020; Zhang et al. 2021), they lead to the economic ecosystem services of this Botanical Garden in particular. In view of the quality of conviviality of the Garden which contains a drinking and peper soup centre, it promotes economic sustainability and boosts the social capital of Akure community at the local level. This aligns with the assertion of Mr. Tola Adeyeba that:

> And also, you can see it provides employment for people here. Look at that *Suya* man outside the gate and the woman in the kitchen with those two boys. Mere looking at them, you can see that they are not actually educated. Not that they really attended a school but it has served as a means of income for them. They might not even have any job now if this place was not [... *available*], and you know that this actually must have save the community and society. When you are not employed, you can just imagine. You become a nuisance to the community and society. So you can see that this place is beneficial to those without jobs.

The socioeconomic ecosystem service described here is an important aspect of social sustainability in ecological urbanism. Social sustainability and social capital support communal wellbeing for all city dwellers. This informs collective responsibility sharing where every citizen contributes to the production stock of the society leading to processes of reducing social inequality and its attendant urban insecurity. Mr. Vincent Iyando equally evaluated the ecosystem services of the Gardens to surpass recreation when he submitted that: "This place is more than a recreation [*site*], it has passed recreation [*for adults*]because there are swings where children can play. They also have siting area apart from this court. There's this social connection with others". The garden is therefore a means of social bonding at the community level. It is also a social space where musical entertainments take place. It affords social mixing of different socioeconomic groups thereby reducing socioeconomic gaps among the various societal groups. These forms of social interventions produced by the Biological Garden could not have been achieved through other means. Through enhancing friendship re-unions, fostering sense of community, meeting new people, and forming new social bonds, the Biological Garden functions as an ecosystem of humans. With the musical displays that feature in the Garden during public holidays, social tensions are eased and a society of happy people with enhanced quality of lives is produced. As these ecosystem services are being obtained, the Government of Ondo State benefits revenue generation that contributes to the fiscal capacity of its urban governance which indirectly boosts that of the entire country. These thoughts were expressed by Mr. Tola Adeyeba who recalled a popular maxim that "little drops of water make the ocean". He argued that "it doesn't matter how minute the income [*revenue*] is to the State Government indirectly, it is adding to the national value". He substantiated his claims with an example of the production industries that produced the drinks sold at the Biological Garden as increasing "the gross revenue of the country" since it was obvious that many visitors and tourists that flock the site everyday translate to more sales. Mr. Vincent Iyando also identified appreciation of nature as a significant centripetal force for the patronage of the Garden, leading to income generation for the operators of its drinks and pepper soup corner and "revenue for the State". Environmental sustainability has been linked with economic sustainability as key components towards achieving the Sustainable Development Goals 2030 Agenda of the United Nations (Leal Filho et al. 2019; Weiland et al. 2021). The Biological Garden and Park, Akure, offers a rich milieu of environmental, economic, and social assets for achieving the Sustainable Development Goals 2030 Agenda of the United Nations at the local, national, and international levels.

The Garden is one of the most extensive urban green infrastructure in Akure and therefore a significant contributor to the micro- and macro-climatic qualities of the city. Mr. Tola Adeyeba refers to this regulating ecosystem service as the most important contribution of the Biological Garden to natural processes in the city to attenuate global warming effects of the reality of climate change. He bared his mind that:

> The trees can reduce that global warming of a thing. And also, they can even detoxify the surroundings. […] You can see this place full of trees and it is purely nature in this place. Like smoke that come from all these cars, the leaves of the tree take in carbon. I don't know what they call it, is it carbon [*di*] oxide and all these emissions that are released from the cars. So, we are safe if it takes them in and gives us oxygen instead.

In addition to the absorption of greenhouse gases including carbon dioxide through carbon sequestration, and most importantly, the metabolism process that supports human life is executed naturally through floral agency that releases oxygen for human respiratory and overall life survival. The contribution of anthropogenic activities to global warming is enormous and urban greenery comes to the rescue through regulating ecosystem services of balancing carbon dioxide content of the atmosphere (Raven and Karley 2006). Considering the emerging Anthropocene, this ecosystem service remains a natural solace for the survival of humans in a world that is constantly ecologically dislocating. In this regard, Gormley (2021) argues that urban landscapes are key components of urban biotic settings and ecosystems. This helps in there imaging and embracing a pre-anthropocenic world of the Holocene. In the thoughts of Gormley (2021, p. 7), "Biotic identity [...] is currently in novel entangling of Holocene adaptations in the ecosystems of the Anthropocene" where "the bidirectional biotic maintenance of organisms and their environs is restructured". To enhance the understanding of the difference, if any, and the areas of conjectures, between the Holocene and the Anthropocene in narrating harzard and disaster stories, Dominey-Howes (2018, p. 12) argues that:

> rather than there being a sharp rupture between the Holocene past and the Anthropocene now, a 'gradualized' transition making the Holocene and the Anthropocene co-existent and one in which humanities impacts on the environment are spread across time obscures the horrifying nature of the Anthropocene and the extraordinary measures required to respond to its challenges.

Therefore, the threats of the contemporary environmental hazards and disasters at global, national, and local scales (Adedeji and Adeboyejo 2011) call for more proactively envisioned ecological strategies for cities. In this regard, Judith Oluwaseun, a female University undergraduate student in the third year, bears similar insight on the regulating ecosystem services of the Garden. "Looking at this place, when you compare those areas where earthquake and other disasters happen, if there were trees there as they are here, it would not have happened", she argued. In her opinion, an important ecosystem service of the Garden is the prevention of destructive effects of earthquakes aside from providing a suitable environment for relaxation due to adequate protection from direct thermal insolation. She premised her assertion on tree covers that "There are trees here, it gives shade to the environment" and "When you are tired, you need a very good place like here to relax". She also noted that: "Here is a very good relaxation centre. All these leaves are beneficial [... *and*] you can just come here to relax and enjoy yourself".

Mr. Tola Adeyeba discussed another highly significant ecosystem service of the Biological Garden on how it contributes to the image of Akure city. He recalled a time in the past "when Ondo [*State*] was ranked third most beautiful State [*in Nigeria*] after Lagos and one other State". To him, "tourist locations affect the image [*of*] the State, and of course, directly or indirectly, the country". He attributed the popularized image of Akure city to the Biological Garden and other ecological elements in the city "like the fountain and the artificial dew Roundabout towards Fiwasaye" area. He drew further insights from how tourist destinations and recreational centres have contributed to the image of Lagos State which has produced centripetal force that attracts multitudes to Lagos. Green spaces promote city image which enhances branding for tourism and offer cities with prospects to boost their attractiveness (Wang 2019). Gilboa et al. (2015) underscore the significance of city branding to attracting good share of the world's tourists among other urban economic growth engines.

Despite these ecosystem services of the Biological Garden and Park, most of the tourists and staff interviewed complained of insect bite which some of them perceive as an insignificant ecosystem disservice compared with the benefits. While explaining the challenges faced at the site, a tourist said "Insect bite is number one. I've been scratching my body since I came" just as another tourist hinted, "there has been mosquito bite since I came here". For Alabi, there was usual fear of snake because of the trees but he was able to overcome that phobia in the Garden, while Tunrayo affirmed the presence of biting insects like mosquitos. Soga and Gaston (2022, p. 55) refer to these ecosystem disservices as "components of human-wildlife conflict". Such conflicts can out rightly preclude or reduce the patronage of urban dwellers that are allergic to insect bites to urban parks and gardens.

2.3 Lekki Conservation Centre (LCC), Lagos, Nigeria

The ecological uniqueness of Lagos as a coastal city with nature-based urbanism in terms of green and blue spaces cannot be overemphasized. This is generally attributed to its exceptionality as having the most diverse green and blue spaces in Yoruba urbanism. As a coastal city in its metropolitan and non-metropolitan settings combined as Lagos State, it parades a plethora of urban aquatic and terrestrial ecosystems that are ubiquitous in its settings. While Lagos is home to numerous wild and cultivated green spaces, Lekki Conservation Centre stands out as a distinctive ecological conservation site that is home to many flora and fauna species in an urban setting. This accounts for its suitability as an urban ecotourism site in the most populous megacity in Nigeria with a population of 14,862,111 people in its metropolitan areas and only second to Kinshasa with a population of 14,970,460 according to the World Population Review (WPR 2021). This section considers the ecosystem services of Lekki Conservation Centre as an urban ecotourism destination.

LCC is a 78 hectares urban park established by a non-governmental organization (NGO), the Nigerian Conservation Fund (NCF) in 1990 as a biodiversity conservation site. It is located in the Lekki Peninsula along the Lagos-Epe Expressway in Eti-Osa Local Government Area of Lagos State, Nigeria. It extends from kilometer 19 to near Okun Ibeju Village very near the Atlantic Ocean (Harrison 2019). The Centre is a recreational site furnished with many installations for vertical and horizontal, passive and active recreations. It has great blends of interactive spaces for human and nonhuman nature andincorporates material constructions with natural elements like trees. Its terrestrial and aquatic splendors in a highly pristine forest milieu of flora and fauna species are enhanced with timber, steel, and hard waring cabling that showcase structural stability, serviceability, and durability. Through these strategies, the LCC presently has "the longest canopy walk in Africa at 410 meters" (Nwokorie and Adeniyi 2021, p. 24). With its Floor Board games, relaxation centre, nature reserve, Family Park, children's playground, wildlife conservation, and auditorium, LCC is home to all visitors. The monkey species, squirrels, birds, tortoise, snakes, crocodiles, tilapia fish pond, and colourful wing birds make the Centre an awesome ecological setting.

These ecosystem qualities of the LCC were extolled by Mr. Ayeni Oladapo when interviewed at the site. As "a lover of nature", there was no other place for him to be as he emphasized further that "the right place to be is to be at the Lekki Conservation Centre". Mr. Ayeni Oladapo saw the site as an enjoyable place that was "fascinating" to him and fulfilled his recreation yearning. The assorted recreation activities afforded by the site projected core values that thrilled him in that primeval ecosystem. His perspective on his experience and appreciation of nature in the site reinforces the understanding of the connectedness of human with the nonhuman nature and inspiresthetruth about the magnitude of values of nature in urban ecotourism sites. This agrees with Fletcher (2015, p. 342) who argues that "ecotourists desire intense, physical, visceral experiences that give them a sense of completion and achievement" and they are mostly urban residents (Ramírez and Santana 2019). Fulfilment of human needs to connect with the natural world is a significant ecosystem service that brings ecological satisfaction. It underlies all passions to access natural landscapes of ecological values in urban settings by residents who are denied the experiences of nature that the countryside affords. The view expressed by Mr. Adegoke Kolade, a Yoruba and tourist, parallels such ecological connection with nature that the site was natural, very cool, entertaining, and pleasurable for him. He visited the site in company of friends and usually made additional friends during his visits thereby extending his social connections and engagements. This attests to the social-ecological significance of ecotourism sites in fostering intimate connection with the natural world that stimulates human wellbeing and multiple health benefits (Karst and Nepal 2022; Soga and Gaston 2020).

Mr. Chinedu Ignatus, a married Igbo man from Imo State, was interviewed at the LCC as a first-time tourist to the site. He was attracted to the site because of his desire to have personal experience of the canopy walkway which he said he had heard a lot about before then. He expressed his satisfaction on the many opportunities to connect with nature at the site. For him, the experience of the Canopy Walkway was not only amazing but the presence of such a recreation facility in Nigeria that is so far the longest and best in Africa beats his imagination. This installation is "impressive for Nigeria", he exclaimed. Obviously, this returnee to Nigeria had never seen something similar to that even outside Nigeria. This is reinforced by his narration that "The experience was nice and from what the Tour guide said, it is the longest Canopy Walkway in Africa, 401 meters, almost half a kilometer". With regular huge crowds at the site, LCC is a cultural melting point for all and sundary, including Nigerians and non-Nigerians alike. Mr. Chinedu Ignatus could not refrain himself from proclaiming that "the closer you are to nature the better" as the site provided him great opportunity "to be away from work" and the associated stress in order to "de-stress", enjoy "the cool breeze, animal sightseeing and the likes". Instead of visiting the villages to experience nature, the site provides a huge disparity between the bustling-hustling urban milieu and a serene 'countryside' in a city setting. Still, he counselled that this experience of nature could be enhanced by providing necessary signages to give sense of direction to tourists exploring the

Canopy Walkway towards going the right directions in return to avoid intercepting in-coming tourists. While appraising the ingenuity of the Canopy Walkway construction, he expressed his concerns on the need for its preemptive maintenance to avoid fatal falls that could result from the deterioration of this recreation element of the site. Despite the extraordinary length of the Canopy Walkway, it is incorporated with resting points for tourists in its natural setting to enhance successful completion of hiking to the end of it without exhaustion. In view of the importance of the ecological and recreational importance of the site, it calls for maintenance and overral management strategies that could support its continuity and ultimately the continued realization of the goals of its establishment. Mr. Chinedu Ignatus equally expected that the gazebos at the site could afford better experience of nature for tourists if they are built around trees. He questioned: "Why not leave the trees as they are and build around them. Imagine that gazebo but with a tree in the centre. Don't you think it will look more natural? It will be more interesting". Experience of nature could be enhanced by incorporating floral elements in the biophilic and bio-inspired design of man-made components of ecotourism recreation sites (Ahmad Sayuti et al. 2019).

The Tree-house at the site presents opportunity to experience nature by climbing. This climbing recreational exercise is a form of Mimi crying the fauna lifestyle in nature at the site and also a means of keeping-fit through a fascinating and hilarious manner. The tree-climbing exercise requires agility which most young adult tourists possess and therefore explore the opportunity as a form of sport, except those that may have phobia for elevations and heights. ForMr. Chinedu Ignatus, he "couldn't climb the Tree-house" as he "saw some boys climbing". To him, he could do the same if he were younger than his present age status. He further pictured, "imagine me climbing up and falling down – and this is New Year", he submitted. His high commendation of this nature conservation in an urban environment is evident, as he identified the LCC as a place "to experience nature, enjoy nature, and see it in its natural, and untouched state". This centripetal attraction to nature makes the site a popular event centre for social gatherings like birthday parties, wedding ceremonies, and corporate meetings in a portion of its outdoor spaces or auditorium. Considering its unique numerous nature attractions, the LCC brings visitors back to nature in a more distinct manner compared to the Ocean Beach. The manner through which the fauna are conserved in their natural habitats with freedom of movement and natural food web contributes greatly to achieving true experience of nature ecologically. This is unlike confined collections of animals in zoological gardens. For instance, the free roaming monkeys could 'interact' with tourists, as Mr. Chinedu Ignatus noted how a monkey reacted to him at the Centre. Marẻchal et al. (2016, p. 1) found evidence that primates in wildlife tourism use different coping strategies with visitors, including "physical avoidance, social support, affiliative, aggressive and displacement behaviours" depending "on a trade-off between perceived risks and potential benefits". The multifarious ecosystem services of LCC also include economic revenue generation, protection of nature from urban development, passive and active recreation, prevention of extinction of species, and achieving ecosystem balance of flora and fauna. Mr. Chinedu Ignatus noted:

> Here, Nature is balanced. It is man that goes into Nature and spoil things. The ecosystem is balanced here. For instance the animals here are free to move and cater for themselves and they're surviving. Imagine if they wiped up this place now, there are some species of animals here that might have gone [*into extinction*]. No one feeds them, and they're living on their own. If we don't touch things they will work on their own.

With such a very rich natural collection of flora and fauna, visitors are brought into experiential contact with nature "in awe of God's creation", as Mr. Chinedu Ignatus puts it. With these huge ecosystem services in an urban place, the site would continue to benefit high patronage of visitors. It would continue to connect human and nonhuman nature together, function as ecosystem outdoor laboratory for learning, and serve as a focal point for urban Lagos. LCC contributes immensely to the micro and macro climatic sustainability of the region, and is continuously admired by those who live in cities and love ecological urbanism.

2.4 Adekunle Fajuyi Park, Ado-Ekiti, Nigeria

Adekunle Fajuyi Park is an ecological memorial located at the city centre of Ado-Ekiti, the capital city of Ekiti State in Southwest Nigeria. Late Lt-Col. Adekunle Fajuyi who lived between 26 June 1926 to 29 July 1966 was a "Military Governor of Western Nigeria who was killed during a military coup in 1966" (Adebanwi 2008, p. 419). Of all the sites discussed in this chapter, the memorial Park epitomizes one of the most critical political events in Nigerian history which is outside the scope of this book (readers who are interested in this subject are referred to Adebanwi 2008, for a detailed account). However, it is important to emphasize that the site affirms the role of nature in documenting urban events at the interface of politics, governance, security, and socio-cultural value-systems. It symbolizes the formation of urban landscapes on the substrate of social capital in Yoruba urbanism, as it contains the burial place and cenotaph of Late Lt-Col. Adekunle Fajuyi, a fallen hero. In view of these considerations, it is apt to refer to this urban park as an ecological-social memorial whose ecosystem services are described in this section. The park has different sections and facilities for adults and children, including green spaces, water elements, and playing equipments. The park has a paddle boat pool which originally had four boats to paddle children and adults but only one remains at the time of this study. Other facilities include a restaurant, conveniences, bar, cinema section, stage, digital lead screen, and VIP lounge. Children park facilities include games, sliders, boat ride, pirate ship, and gyroscope.

Mr. Julius Anifowose, a Christian, Bachelor's degree holder, and 43 years old Yoruba of Ibadan origin was a tourist with repeated visits to the Park when interviewed there. To him, the park is a relaxation point to "have some good time" because of its quietness, aesthetic values, coolness, and peaceful serenity compared with the buzzing and hustling city. The park had always served as a place to socialize and "hang out and enjoy the fun facilities" there. He recalled the memorial significance of the Park and why it was named after Late Lt-Col. Fajuyi. He reiterated that the site was reminiscent of this "military man" that "served under the British [*colonial*] government as at the time before he became the first military governor of the [*defunct*] Western Region [*of Nigeria*]. To immortalize him, "most western States had a street named after him, even in Ibadan […], there is Adekunle Fajuyi Cantonment at Akobo Road, around Alaakia Estate", Mr. Julius Anifowose submitted. In his view, the current public-private partnership management strategy of the park made it to be wearing a better look compared with Ikogosi Warm Spring resort tourists' destination which is also located in Ekiti State. The park appears to be inscribed in the conscious and sub-consciousness of the citizens. It is the first place that "comes to mind" when desires to "hang out in a natural environment" pump up. This is aside from contributing to the financial base of Ekiti State as a source of internally generated revenue for the government through gate fees and rentage of spaces for social events. Until its rebranding as an amusement park, the Adekunle Fajuyi Park appeared to be a breeding ground for social vices and disservices like prostitution, secret cult activities, and dealings in hard drugs because of its contextual location being surrounded with many beer parlours. Another interviewee, Mr. Ipadeola Adetunji gave further insights on the park. Mr. Ipadeola Adetunji, is a 44 years old man who holds a Bachelor of Technology degree, married, Yoruba, and an indigene of Oyo State was interviewed as a staff at the Park. He works with a bigger recreation business in Ekiti State that renovated and rebranded the park and manages it. According to him, the ecosystem services of the park include a space and place for socio-cultural services like reuniting families, bonding relationships, and securing the social capital of the society through recreation. The park also serves at the space for the annual Army Remembrance Day in Ekiti State to commemorate the roles of fallen heroes in the Nigerian military. For this purpose, the park holds the statue of an unknown soldier in addition to that of Late Lt-Col. Adekunle Fajuyi.

The cultural ecosystem services of urban memorial landscapes are evident in these narratives. In this scenario, the Adekunle Fajuyi Park, Ado-Ekiti, embodies intangible heritage of heroic patriotism in urban governance. Eliasson et al. (2018) argue that cultural ecosystem services have roles to play in heritage planning and particularly material planning for intangible heritage. They applied the concept of ecosystem services to link "cultural heritage, place identity, and aesthetic and existential values" (p. 44). They link place identity and place memorial as being critical to collective memory as cultural ecosystem service. The integration of culture and nature in cultural ecosystem services makes the concept valuable for assessing urban heritage landscape to understand critical aspects of intangible heritages in urban governance. Since a place is a space with identity (Carmona et al. 2003) the Adekunle Fajuyi Park, Ado-Ekiti, fosters emotional connection with the past political events in Yoruba urban spaces and how such events are ecologically documented and resonated in the mind of tourists. Till (2005) argues that memorials describe pasts and futures of a cultural group through the spaces and times of a city and the city itself is a place of social memory. Thus, the Park has both national and natural values in socio-ecological sense.

2.5 Muri Okunola Park, Lagos, Nigeria

Muri Okunola Park is one of the most popular public green spaces in metropolitan Lagos. It is a green memoryscape located along Adeyemo Alakija Street, Victoria Island, Lagos, near the Nigerian Law School Campus. It was established in 2008 in honour of a late hero of the Nigerian Judiciary, Justice Muri Okunola. The Park showcases a rich combination of memory, landscape, urban ecology, and social justice. Therefore, its ecosystem services are multifarious. It is a truly public ecological place where gate fees are not collected from visitors.

Miss Deborah Adebayo, a Bachelor's degree holder in Political Science and Yoruba of Osun State origin but lived in Lagos Island was at the age of 26 years when interviewed at the Park. She had been a regular visitor to the Park for the past 2 years prior to the interview. Her very regular visits to the Park were motivated by its close proximity to her residence at the Lagos Island. Residential proximity to urban public green spaces enhances access to and use of such spaces and therefore exploring their ecosystem services. A short walking distance of an average of 15 min from home to the Park was an impetus for repeated visits for Deborah Adebayo. She narrated her initial contact with the Park and how she had been attracted to the site since then. In her narrative account, she revealed the following.

> Two years ago, my friends and I wanted to go for a picnic. We had been anticipating going for one, so we were checking for different parks around the Island because this is where most of us live and then we now found this park and it was really close to my house, and when we came I really enjoyed my time here, the environment and all. So since then I just found myself coming more often.

The park not only supplied her ecosystem service demand, it became a moderating entity for her social life, as she explained:

> I usually come to the park whenever I just need to relax and whenever they get me angry at home. Instead of getting pissed, I just come here to feel better or whenever I want to think about something. […] I also come here to take pictures once in a while.

Urban green infrastructure can mediate in social disruptions. In the case described, the Park serves as antidote for negative emotions like anger, frustration and resentment. Burley (2018) discovers that the presence of urban green infrastructure is associated with an increase in positive emotions like feelings of happiness and reduction of negative emotions. Little wonder that Deborah developed place attachment to the Park as a major component of her lifestyle. Her repeated visits were propelled by her passionate love for nature in exchange for boring home interiors and bustling urban life of Lagos metropolis. "You know how stressful Lagos can be, so after work I come here to relax and forget my sorrows", she explained. The Park serves as socio-ecological space for meeting new people, wedding ceremonies, photo shoots, small-group parties, dinners, get-togethers, festivals, recreation, and relaxation. Using the Park for ceremonies like wedding also attract rental fees just as the photographers that use the Park for photo shooting "pay hourly". This adds to the fiscal base, and therefore a socio-economic ecosystem service, as internally generated revenue of the Lagos State Government that operates the Park through the Lagos State Parks and Gardens Agency (LASPARK). To make strong case for the recreational benefits of the park in terms of human health and wellbeing as very crucial ecosystem services in a place like Lagos, Deborah expatiated that the Park:

> Helps people to relax. Like I said, Lagos is very stressful. Everybody knows, once you come down from those buses, the loud noises, [*bus*] conductors disturbing you. Lagos is stressful and everyone needs to rest for good health. So, the health benefit is that if you come here you are able to relax. At least there's somewhere in Lagos where you can relax [*and*] forget your sorrows [...] and feel calm.

The importance of recreational ecosystem services of green spaces in such a busy urban landscape like Lagos cannot be overemphasized as a nature-based solution to unhealthy urban sonic lifestyle. Her narration presents a contrast between built and unbuilt landscapes of cities. While the city life is chaotic, she identified the Park as "a very beautiful place" where visitors "can just appreciate the work of God, nature" and balance "social life" in a serene environment with very high air quality compared with the other city spaces. As a place that enhaces filial social capital, "It is important for recreation like families coming together to have fun and just be together, like it can be a gathering of families or social groups". This made some couples to visit the park frequently, just sit down, "have a quiet time and talk" as "people that just love nature". The park offers opportunity for meeting new people, and establishes new social connections. These ecosystem services are testimonies of the power of nature in cementing human social relationships and fostering family bonding and togetherness. Deborah argued that couples and families need to go out of their "comfort zone" because "It's not every time you need to have dinner on your dining table at home, so come out", she submitted. She added that she "usually looks forward to doing that too". She liked to "see families with their children and everyone is there, they're just having fun. Yeah, and then couples too. It's good for you to visit and have fun", expressing her desire for the future married life. The park was considered very safe for these activities especially in the day time because of its being a defensible space with public characters including safety and security surveillance.

The expansiveness of the park makes it suitable for healthy walking or strolling in a serene environment. The serenity is ensured by disallowing unwanted noise from music and other sources and all visitors are normally equipped with personal headsets to avoid open music that could disturb other visitors. For Deborah Adebayo, just like "people that love nature", the Park was not only serene but also naturally "quite colourful" and green and "anyone that likes nature will love this place", she added. The Park made her "feel relaxed and safe anytime" she was there, her favourite visiting time being evenings for relaxation "after a stressful day" except for picnics that could be in the morning, afternoon, or evening. The installation of colourful artificial lighting system in the Park makes it suitable for recreation and social activities in late evenings as a "really nice place to be", as she sketched a narrative about a marriage proposal event she once witnessed in the park:

> Yeah, it was so beautiful and the girl had no idea. It was actually decorated for that purpose. So, that's one of the most special things or events that I've seen happened at this park. It was actually a very beautiful experience and I mean, it was really nice and after the proposal, they took pictures and had fun.

The park had also contributed in no small way to her social life which she believed was a divine benevolence. She recalled that she met one of her very good friends named Seun in the Park through which they became close and maintained the contact – "Yeah, here at the park and then we just clicked and now we're very very close", she affirmed. She made allusion to her visit to the Dubai Miracle Garden to draw comparative lessons with the Muri Okunola Park as she recalled:

> That place was so fine. There were a lot of flowers, I mean, and we even paid so heavily for it. So even if they can do something like that here where there are really many...like now, there the flowers [*plants*] were made into people [*human figures*], a castle, made into big teddies and then it was a very nice place to take pictures, like I took over a thousand pictures in that location alone. Ah, it was really nice, I'm serious. So, if they could actually do something like that, maybe not the whole place, but then that kind of thing will need a lot of maintenance.

In her ecosystem service assessment of the Dubai Miracle Garden, no amount of gate fee for visitors could have been too much to have access to such a awesome experience of nature. Therefore, she desired that such nature endowed space could be replicated at the Muri Okunola Park where visitors may even have to pay entrance fees "because of maintenance, that would make the park more beautiful". Such plant topiary is an aesthetic horticultural practice of shaping foliages and requires very regular trimming maintenance in view of the vegetative dynamism of floral parts of plants. In a similar futuristic sense, Deborah would also desire an owner-occupied built residence reminiscent of park environment in terms of green spaces and general nature embellishment. She drew this vision from her understanding of the ecological significance of trees and green spaces in human metabolism as supportive ecosystem services that nature elements:

> Make the air cleaner; we're breathing in oxygen and releasing carbon dioxide, the trees are releasing oxygen and taking in the carbon dioxide. So yes, I would love to have green areas in our house as you said. And having a park close to my house also means that the place [*and*] the environment will be cleaner and so yeah, when it comes to building [...] I think, this park, it's one of the reasons I love where I stay currently.

With such actively expressed love for nature, trees, "beautiful flowers" and green areas generally, Deborah demonstrated a huge biophilic connection to nonhuman nature in a real urban ecological sense.

Despite these ecosystem services of the Park, it creates a negative image for Lagos Island as a place for homeless people that used the Park at its gateas begging space. These homeless beggers constitute social disorder to visitors to the Park because "they can be so pushy, that they keep on following you till sometimes some people scream at them", as Miss Doris Adeeyoexplicated. This kind of socio-spatial scenario "actually makes people uncomfortable" where someone is just following visitors to the park in questionable and irritating manners. Also, the presence of the

Park increased the rental values of real estates in its neighbourhoods. This kind of incremental effect can mean ecosystem disservice to urbanites who are not able to afford rent of such properties or land resources. Others are not able to secure any interest (that is, bundle of rights) in real estate in the park's neighbourhood. Miss Doris Adeeyo illuminated on this as she submitted that:

> Of course, I mean, yes it would surely increase. In fact, anytime you want to buy these lands or buy a house or anything, they usually write it there if there's a recreational park in the area. So, yeah, it actually increases economic value and it just gives that environment that…just like what I'm saying about myself, after finding out how this place is. If it was far from my house, I may not be visiting here as often as I usually did [sic].

Similarly, the literature is replete with the relationship between real estate values and proximity to public green spaces, parks, and gardens where these natural elements of urban landscapes increase real estate values across continents and in decades of time (Crompton 2005; Voicu and Been 2008; Jim and Chen 2010; Wu et al. 2017; Crompton and Nicholls 2019;). Through a wide review of the literature on this theme and a case study, d'Acci (2019, p. 71) confirmed that there is an intricate connection between the characteristics of the site of a dwelling and the "decision processes" of households "when choosing among alternative dwellings". Nesticò et al. (2020) linked access to carbon sequestration of tree canopy area as ecosystem services to increased real estate values while Chen et al. (2021, p. 1) found that "green spaces do cause house prices increases". These narratives are exemplar evidences that though urban public parks and green infrastructure generally provide diverse ecosystem services, they are nonetheless sources of some ecosystem disservices that affect different groups of urban dwellers at different degrees.

2.6 University Campus Green Spaces in Yoruba Cities

Yoruba cities can be aptly described as a good blend of 'town and gown' with most of the cities as home to large University campuses that constitute major land uses. The green spaces of these University campuses not only contribute immensely to the ecological framework of their host cities but a good number of the Universities are cities in their own rights with highly rich ecological networks. According to Le Corbusier in 1936, the […] campus is a world in itself, a temporary paradise, a pleasant stage in life (Campos 2010). As a kind of city in microcosm, the campus provides a good balance among the three magnets of the town, the country and the people (Howard 2010). The plethora of green spaces in University campus in Yoruba cities include parks, gardens, conservation areas, grass lawns, ornamental greeneries, and recreation fields. In addition, hydrological elements that are rich in ecological biodiversity in the campus urbanism include lagoons, dams, and rivers. The ecological settings and ecosystem services of some of the University campuses are discussed in this section, especially botanical gardens.

2.6.1 University of Ibadan Botanical Gardens, Ibadan, Nigeria

The University of Ibadan Botanical Gardens, Ibadan, were established the same year 1948 when the University was established as a College of the University of London. University of Ibadan is a campus city built on 2550 acres of land (Nwachukwu et al. 2021) with a total students' population of 32,228 and 2736 staff, both academic and non-academic staff members (Adebayo 2019). The University campus has numerous institutes, colleges, and faculties, accommodation areas for all categories of students, and staff quarters. It has two secondary schools, one primary school, sports stadium, conference centre, water dam, zoological garden, and botanical gardens – indeed, a campus city. The Botanical Gardens cover a total land mass of over 100 acres with assorted trees, shrubs, and herbs. It has a number of Gardens including Water and Bog for plants requiring waterlogged environment, Medicinal Plants section, and Ornamental Plants for environmental beautification shrubs. Other gardens include Rock Garden section with Xerophytes that require less water to survive, Rose, Children, Conservation Forest, Open Field, and Arboretum Garden for different species of trees like Iroko (*Milicia excelsa*, F. Moraceae) and Obeche (*Triplochiton scleroxylon*, F. Malvaceae). The Rose (F. Rosaceae) Garden contains pink, yellow, white, purple, and local bush roses. The Gardens mostly consist of insitu plants and less exsitu plants. The insitu plants grew naturally on the site and were therefore indigenous and the exsitu plants were exotic, being imported from foreign countries outside Nigeria. The tags on the plants, most especially trees, showed their common name, botanical name, family name, and origin. Mr. Osayinpeju Tolulope, who holds a Master's degree in Computation Environmental Biology was a Senior Field Assistant of the University of Ibadan (UI) Botanical Gardens and he was 32 years old when interviewed at the Gardens. He was a Yoruba man from Ogun State and had worked in the Gardens for up to 7 years. Mr. Osayinpeju Tolulope explained that the Herbarium of the Gardens has over 26,000 flowering plants collections along with Bryophytes. Other facilities in the Gardens include garden furnitures and picnic gazebos that provide recreation spaces in natural setting, and public conveniences.

The primary ecosystem services of the Gardens are teaching and research purposes while the secondary ecosystem service is recreation. The need to serve recreation purpose led to the incorporation of gazebos, snacks bar, and garden furniture where people can "sit down, have fun, and talk", as Mr. Osayinpeju Tolulope explained. The teaching ecosystem services are being achieved in the following way as explained by him:

> Students come around very well for educational visitation. Those are the Faculty of Pharmacy, Veterinary [*students*] – at least most departments. I know of pharmacy students, chemistry students, pharmacognosis, and the likes that make use of plant parts for their projects, and there are aspects in botany that when you come around you do something like physical features of plants.

The conventional status of the University with many plant-based courses requires the Botanical Gardens for teaching and research. The Gardens support zoological research on the types of birds in the environment. Mr. Osayinpeju Tolulope illustrated this claim as follows:

> There was a student I know. The lady comes around, she spreads her net in the garden, because you know birds like trees, and she wants to know the kind of species of bird that are actually in this environment. So she spreads her net. Anytime the birds fly, they will feel like they want to rest on the particular tree, so in the cause of their trying to migrate to the tree the net caught them. They try to force themselves out of that net.

The nursery Garden also produces seedlings for procurement by the members of thepublic. That makes the Garden to be significant for economic ecosystem service, thereby contributing to the internally generated revenue of the University. These could be ornamental plants that are usually chubby, including some shrubs that look like trees. The recreational ecosystem services of the Gardens have been expanded to include activities like picnics, family get-together, birthday parties, and wedding ceremonies at the open field. These activities also take place under tree canopies which also attract payment to the Gardens' Management. According to Mr. Ogunsanya Babatola, "large places [*spaces*] under the trees" are used for "family parties and the likes" under the tree canopies. He further added that:

> We had a seminar [*programme*] here last week Saturday. They had an attendance of about 200 people under the trees. It was shady, so at least there was no need to get canopies.

The choice of the Botanical Gardens for social gatherings was usually facilitated by persuation and preference for green environments compared with indoor event centres. Mr. Osayinpeju Tolulopecited an example of a wedding ceremony that took place in the Gardens about 4 weeks earlier that involved a bride and bridegroom that went to Nigeria from the United States of America. Obviously, it was not that they could not afford indoor event halls and the likes but they preferred a green environment of nature. Visitors also pay One-hundred naira per person to procure access tickets to the Gardens. Mr. Osayinpeju Tolulope emphasized that:

> The incomes are made for IGR (Internally Generated Revenue) because when they receive payments, all those plants are sold with payment receipts. Like the way you are coming in, you should have gotten a ticket there, it is 100 naira per person. So those tickets that you get, the IGR [*Office*] comes around and receive the money, [*and*] they remit to the University's account.

As a source of agricultural seedlings through the Nursery Garden, the Gardens support tree crop production in the region. Different propagation methods of sexual and asexual approaches were employed. The asexual propagation uses plant parts like stem cutting, budding, and graphing while the sexual propagation make use of the seed. The plants also supply all kinds of foods, building and furniture materials. As a "house of plants", the Botanical Gardens also "make us understand what nature looks like", Mr. Osayinpeju Tolulope submitted. Many plants in the Gardens also serve ethno-medicinal purposes. They are "very good for our health, but it is for you to identify them", he emphasized as he pointed to Cedrela odorata (F. Meliacea) as an example. The plant has "gum that pharmaceutical companies use in conjoining drugs".

The regulating ecosystem services of the Gardens contribute greatly to environmental sustainability. It is an outdoor laboratory "where nature can be really explainable", Mr. Osayinpeju Tolulope noted. The depletion of ozone layer through environmental pollution of carbon monoxide from automobiles, bush and grass burning, among others, that erupt into the air is regulated through green spaces of the Gardens. The green spaces also act as air-purifiers which account for the outdoor environmental air quality of the Gardens. The Gardens serve as nature conservation area against deforestation with some species of trees already going into extinction in the forests of the Yoruba region whereas such trees are still present in the Gardens with some up to 70yearsof age. The rate of deforestation in the region had been exacerbated because of increased rate of urbanisation requiring forest timber resources for building construction and furniture making.

The recreational services of the Gardens in supporting the human 'total being' cannot be overstressed. Mr. Ayokunbi Amos was single in his mid-twenties and a Yoruba man who resided in Ibadan but an indigene of Osun State from Oriade Local Government Area. He was a public speaker and business consultant and had visited the Gardens up to six times as at the moment of interview in the Gardens. He appraised the quality of the Gardens as a meditation centre through inspiration and what he called "experience of nature". He described the connection of this "experience" with his productive mental outputs according to the following further details:

> Being a business consultant, whenever I want to just ideate and think, whenever I want to meditate, I go to the botanical garden because of the experience of nature, the way the plants are tilting to the wind, the way the birds are pecking on the trees is an amazing sight and It's a deep well of inspiration for me. So that is one of the mental benefits of [...*the*] Garden. It's so serene, it's an environment where you could think, you could ideate, you could put your thoughts together.

Mr. Ayokunbi Amos equally explained the roles of the Gardens' plants in human metabolism process that "plants are a great source of Oxygen because they exhale Oxygen and [*while*] we inhale Oxygen and we exhale Carbon IV oxide" which make plants very crucial to human survival. Chinaza was 20 years old and a first-time visitor to the Gardens. He was a student of the Federal University of Technology, Owerri who originated from Anambra but lived in Lagos. He was undergoing Students' Industrial Work Experience Scheme in Ibadan when interviewed in the Gardens. These advantages of the Gardens were beautifully captured by her:

> Yes, it's cool, quiet, based on its location. So, it's somewhere you can come and reflect on things. Life is a hustle, life is a struggle, life is hard, so you can come here to get relaxed, like calm your mind down from the stress of life from the outside world just take a break off stress, relax or if, maybe you're someone who has artistic view and all, yeah, you can come here to get more ideas and reflect like, being a writer, it can be a place you can come to relax and get more ideas and all.

The Gardens provide territorial freedom for all users, allowing some temporary personalization of public space for productive meditation away from the sonic hullabaloos of city centres. Isaiah was at the age of 24 years and an indigene of Lagos State. He was single, a Christian, obtained a Diploma Certificate, and lived in Lagos when interviewed in the Gardens during his third-time visit. He was a creative writer that explored the territorial benefits in the Gardens for fecund mental praxis as he explained that:

> I am a creative writer like I said earlier. So sometimes I need my own space, I can spend time alone to reminisce, to meditate a lot especially in places that have to do with nature, are some greens, some environmental factors, some elements of the nature for me to just feel very comfortable to think deep and to think far and wide, how I can actually put my thought process into words.

To Isaiah, the ecosystem services of the Gardens are all-encompassing. He pointed out the physical exercise benefits to include walking, dissipating excess body calories, and associated health benefits in the body metabolism. He explained the ecosystem services of the Gardens further from the perspective of public gatherings:

> If people are actually at the Botanical Gardens to have a public gathering, this is supposed to actually give some excitement; but then you can actually still exercise from the act of your talking, playing, laughing, discussing. For mental [*benefits*]; people like me that actually come to the botanical garden to have creative thinking, it helps me think in the direction of where I want to think. It actually helps me see things in a different light and I can actually say I related to my inner self better when I'm in the garden or in the nature zone. So, I will say, that is mental for me. That is the kind of mental benefits and physical benefits that I've mentioned here.

The cultural ecosystem services of the Gardens described here are enormous in terms of productive psychological wellbeing and mental health benefits. They confirm evidences in the literature that contact with green spaces provides nonmaterial ecosystem services that improve mental health of humans. In a nuanced review of these evidences, Aerts et al. (2018, p. 5) found that the magnitude of contact with green spaces determines "the magnitude of their positive health effects". The authors link biodiversity, natural capital, human health and wellbeing, and ecosystem services as constituting a complete ecological system. They situate this framework within the "biodiversity hypothesis" that "exposure to biodiversity improves the immune system by regulating the species composition of the human microbiome" (p. 8). Through a wide and deep scoping review framework of numerous research works on green spaces, Callaghan et al. (2021, p. 179) discover that "green spaces are associated with positive mental health outcomes". These evidences have proved that the positive psychological benefits of green spaces are important cultural ecosystem services in all human populations.

These psychological cultural ecosystem services are confirmed further at the University of Ibadan Botanical Gardens. Mr. Akanbi Oluwagbemileke was a 23 years old Yoruba from Ikere-Ekiti but lived in Ibadan and was single with a Bachelor of Technology degree in Statistics when interviewed in the Gardens. He extolled the spatial magnificience of the Gardens as being a great place for exercise around "and keep[*ing*] fit". Also, Mr. Alaba Adedapo was a married Yoruba man from Ekiti State and also a music instrumentalist that lived in Ibadan when interviewed in the Gardens. He discussed the psychosomatic cultural ecosystem services he derived from visit to the Gardens in a mental metabolism manner between the human mind and body:

> It calms the nerves, it calms your reasoning, you know, you let go of all these stress of a thing. You find a place where you can relax your brain, and then have a very cool environment, very conducive environment and at the end of the day, while leaving the place, you'll know that something has come into you and some things are out of you.

How these ecosystem services are interwoven was discussed by another visitor to the Gardens, named Mr. Bibire Temitope, at 23 years old, single Yoruba, and an undergraduate student of the University of Ibadan who hails from Ondo State. According to him, the Gardens meet the needs of people who "just want to chill and relax" and "to just meditate" in view of "very beautiful scenery with a lot [*of*] peace and quietness"away from distractions. These natural qualities make the Gardens suitable for being alone and "have a quiet time" while other "people can also come around, have fun together, create a little event like a picnic or something", he added. The spatial expansiveness of the Gardens affords individual and group activities without constituting any interference and disturbance to other visitors nor vice-versa. He captured this freedom of use and its connection to human-nonhuman ecological relationships succinctly as follows:

> In terms of a green area like the Gardens, I mean you can just make that place a routine and try and get yourself there. I can just go to the Gardens anytime, whenever I want to be in my own thought or something and I want to be out of everybody's space. It is a good place to just go and chill and relax without anybody disturbing me, and people can just find time to get to know each other that way in the Gardens. I mean the garden is a very beautiful place, there are plants and all of that. You can also just get there to see how beautiful the scenery is, and how beautiful our environment is just to understand that, okay, these plants are also part of us.

His assertion reinforces the innate human connection to the nonhuman natural world. Such attraction to and connection with nature are very important ecosystem services for Oluwakunmi, another 23 years old lady visitor. She was a final-year student of the University of Ibadan and an indigene of Ibadan, single, and a Christian. She highlighted her love for flowering plants, and the kind of connection with nature she derived from the Gardens because "there are a lot of places" that visitors "can just sit down" and "it's just you, the trees", and "the flowering" plants. To her, keeping fit could be easily achieved by visiting the Gardens, as she exclaimed:

> You'll lose weight, you'll feel good like body-mind connection. Once your mind feels good, once your body feels good your mind feels good, the body feels good. So, when you're with nature, you'll be able to achieve that faster than you can inside your room.

She visits the Gardens for "mental benefits [...] mostly whenever [*she*] needs to think well and to relax", and generally to derive benefits of "being around nature", having a feeling of natural environment, as she confessed, "I love the cooling effects". These experiences of nature for visitors in the Gardens are highly critical cultural ecosystem services in view of the hot tropical climate of the region. Within this regional climatic context, green spaces with tree canopies are imbued with comfortable micro-climatic mileau especially on sunny days. This characteristic climate of the region makes a good number of visitors to schedule their visits to evening periods. For instance,

Mr.Akinyooye Matthew, who gave a range of his age to be between 25 and 30 years, was a postgraduate student in the University of Ibadan, single and from Ondo State. He was interviewed in the Gardens and he submitted as follows:

> My frequency of visit to this garden has always been often and the time of visit, mostly it has been towards the evening when the environment is cool and serene. My purpose [*of visit is*] actually because I am an environmentalist. I like observing the environment, both the plants and the flowers, there are a lot I learnt from nature and one thing I can clearly say about my love for nature is that it gives me a sense of joy and reduces stress and in a nutshell, I can say I am a lover of nature.

Lifetime love for nature has been found to be a propelling force for nature-based tourism in green spaces and conservation areas (Hosaka et al. 2018; de Kleyn et al. 2020). It contributes largely to the appreciation of nature and how visitors to botanical gardens recognize an affective bond that is innately ecological between human and nonhuman nature. These narratives explain the different mechanisms with which humans value, negotiate, and navigate their ways through in urban settings that are generally non-natural to achieve ecological satisfaction. Through these lenses, ecological urbanism comes to the rescue of city dwellers to reconnect them back to nature from which they have been alienated. This was due to the earlier mis-interpretation of cities as built forms instead of intertwined ecosystems of humans and nonhumans merely 'sprinkled' with essentially unavoidable built forms, if any, that are required by the duos to function ecologically. These arguments lead back to the larger issues of how cities should be configured in terms of nature – either as nature in the city or nature as city. Despite all the numerous ecosystem services of the Gardens, the presence of tiny biting insects is a challenging ecosystem disservice.

2.6.2 Federal University of Agriculture Abeokuta Botanical Garden, Abeokuta, Nigeria

The Botanical Garden of the Federal University of Agriculture Abeokuta (FUNAAB) is unique in many senses. In the first instance, it has urban location with obvious ecological contribution to the city's ecosystem. Secondly, it is the only pure and applied University botanical garden in Yoruba cities with pure horticultural and agricultural focuses. Thirdly, the botanical garden is young in age compared with the high quality of its plant collections and the only pure and applied botanical garden in the region. The University, FUNAAB, is a specialized institution established on 1st January, 1988 on about one-hundred thousand acres of land in the North-East end of the Yoruba city, Abeokuta. The Garden has different sections like Arboretum, Nursery, Vegetables, Ornamental, Fruits, Medicinal, and Toxic Sections. It is also incorporated with a Children Park/Section. It is transversed by a stream of water which contributes greatly to its ecological quality and supplies the water for artificial irrigation of the Garden. The Toxic Section contains toxic, harmful, and poisonous plants for various uses. It is operated mainly for research purposes as an educational ecosystem service in collaboration with the Veterinary Medicine Department of the University. According to the latest floral assessment of the Garden in 2016 as given by the key informant Mr. Oluwatimileyin Isawumi, it contains about 210 plant varieties. The multifarious ecosystem services of the Garden is evident in its holding of medicinal herbs, varieties of trees in its arboretum, and assorted vegetables in its plant collections. These were enhanced by ecological and topographical qualities of the site and should have been confirmed through floral survey at the planning stage. Trees in the Garden include Monkey-ear or Elephant-ear tree (*Enterolobium cyclocarpum*, F. Legumiunosae-Mimosoideae), *Dogonyaro*, Neem (*Azadirachta indica*, F. Meliaceae) and Teak (*Tectona grandis*, F. Verbenaceae). They provide adequate shading canopies for the Nursery Section and the topography enhances its hydrological performance. Through taxonomist approach, the Garden plants were properly labelled for immediate education of students and general visitors.

Mr. Oluwatimileyin Isawumi, a Yoruba man, who is single, at the age of 27 years, and holds a Bachelor's degree in Pure and Applied Botany from FUNAAB and a Master's degree from the University of Ibadan. Hewas a Supervisor and key informant interviewed in the Botanical Garden to unpack its ecosystem services as a significant ecological milieu in the Abeokuta city. Mr. Oluwasanmi Adewumi has a long period of connection with the garden. The garden was an outdoor laboratory for his undergraduate pragramme and thereafter employed to serve as Supervisor which made him to demonstrate deep knowledge of the value-system of the garden. Even in-between these times, he frequented the Botanical Garden passionately during his National Youth Service and Master's programme to assist in guiding the FUNAAB students for practicals. To him, urban botanical gardens are very crucial to curating "different types of plants that perform different functions for different purposes". He explained that while University-based botanical gardens are primarily meant for research and recreation, gardens of various categories exist outside the University campus. There are green spaces in residential neighbourhoods and numerous horticultural gardens in the Abeokuta city. The horticultural gardens are commercial outlets that sell plant seedlings for various landscaping purposes. The FUNAAB Botanical Garden also serve the people outside the University community for procurement of genetically-improved plant seedlings, education, and recreation like picnics and get-togethers in the Arboretum. Therefore, it is a source of income to the University Management and adds to its economic base in some significant ways. Mr. Oluwasanmi Adewumi submitted that "those people that visit the garden, [*they*] don't visit for free", they pay entrance fees, "people that come for ornamentals" also pay like "the people that come for picnics". The year-four students of the bachelor programme in Botany were regular users of the Garden for their final-year academic research project in the University. Similarly, the postgraduate students make use of the Garden for their researches. The Garden is reputable for its socio-cultural ecosystem services as a relaxation and recreation centre. For instance, Mr. Oluwasanmi Adewumi submitted that:

> Some people come in here to relax. See how cool this place is. Some people come here and just lay down on this bamboo seat and just sleep and enjoy the cool and fresh air [*and*] the garden should be a place where people should actually come [... *to*] know about plants.

In addition, the ethnobotanic ecosystem service value of the Garden is obvious. For instance, it contains White Teak (*Gmelina arborea*, F. Lamiaceae) which is a highly medicinal tropical plant that is also valued for its pharmacological significance in other places like India. In an extensive review of the medicinal and pharmacognostic ecosystem service of this plant, Warrier et al. (2020, p.5) discover that it contains "sixty-nine phytochemicals" with majority of them already categorized for their pharmacological qualities. These include "wound-healing and antidiarrheal properties", aside from activities like "anti-oxidant, anti-diabetic, anti-inflamatory, antiulcer, analgesic, anti-nociceptive, anticancer". From their rigorous review of the literature, they provide a detail provisional ecosystem service analysis of the plant as follow:

> The whole plant is used in medicine. It is astringent, bitter, digestive, cardiotonic, diuretic, laxative, and pulmonary and nervine tonic. It improves digestion, memory, helps overcome giddiness and is useful in burning sensation, fever, thirst, emaciation, heart diseases, nervous disorders and piles. The roots are acrid, bitter-sweet in taste, stomachic, tonic, laxative, galactagogue and antihelmintic. The flowers are sweet, refrigerant, bitter, astringent, aphrodisiac, trichogenous, alterant and tonic. Fruits are edible and also used for promoting hair growth (p.5).

African oil palm (*Elaeis guineensis*, F. Arecaceae), both trees and seedlings, was also among the numerous plants with multifarious ecosystem services in the Garden. It is also perhaps the most common plant in some urban and mostly periurban areas in Yoruba cities especially areas with sprawling urbanisation with deep incursion into the nearby forests. It forms important urban ecological belt in some cities where it exists as an urban plantation or in unbuildable waterlogged zones of Yoruba cities or along urban waterscapes. Sometimes, it is also planted for ornamental purposes in residential neighbourhoods or urban public spaces related to the administration of agricultural practices. All parts of Oil palm trees are important for one ecosystem service or the other. The trees are significant habitats for urban ecosystem populations of birds, rodents, and other animals including mammals as contributor of supporting ecosystem services. Its fibrous root system makes it important for regulating ecosystem services in enhancing soil cohesion and inhibiting the inception of soil erosion due to runoff process. When the trees occur in appreciable number in groups, they have capacity to contribute meaningfully to urban carbon sink, act as air purifier, and regulate microclimate for outdoor use especially during hot periods and peak urban heat. As supplier of provisioning ecosystem services, the trunks of fully grown oil palm trees have economic significance. They could be sawn to support carpentary works in residential buildings as local incentives on lowcost home delivery. The trees are sources of palm wine which is a local traditional drink. In urban and suburban areas of Yoruba cities, the fronds of oil palm trees are processed into brooms and baskets while fish traps are also produced from them in urban riverine areas. Despite all these, there are huge economic benefits of palm oil fruits through multiple provisioning ecosystem services. The processed fruits are sources of palm oils while the kernel meat can be processed into livestock feed. The fruits also contain palm kernel nuts that are processed into different industrial types of oils. The byproducts of palm kernel cake (PKC) are used as livestock feeds and the kernel shell is useful for steam boilers. Aweto (2001) discovered that African oil palm also contributes immensely to the floral quality, biodiversity, and soil fertility in the urban fringe of Ibadan, another Yoruba city located about 85 km from Abeokuta.

Furthermore, the Ornamental section of the FUNAAB Botanical Garden is enriched with floral collections like Ixora coccinea (F. Rubiacea), Oyster plant (*Rhoeo discolor*, F. Commelinaceae), Arbor vitae (*Thuja occidentalis*, F. Cupressaceae), White buttercup (*Turnera ulmifolia* and *Turnera subulate*, F. Passifloraceae). Others are Agave sisalana (F. Agavaceae), and Euphorbia tithymaloides (*Pedilanthus tithymaloides*, F. Euphorbiaceae). These plantscontribute greatly to the aesthetic quality of the environment aside from their supporting and regulating ecosystem services. They are ornamental plants that are propagated by either seeds, sucker, stem cutting, or other specialized forms like air-layering also referred to as marcotting, or simple layering. These processes are supported by the top soil of the Garden and application of manures, especially in the nurseries. The choice of ornamental plants for different ecosystem services could be based on their beautiful flowers, life-cycle, foliage system, colour, odour, or those that double as medicinal plants.

The Garden suffers from some ecosystem disservices through its food web. These include predatory insects and larva that affect the vegetables section, roaming reptiles like snakes that are scary to visitors, and change in season that leads to fire incident during the dry season. For instance, Mr. Oluwatimileyin recalled a fire incident that took place in the Garden sometime and "almost all the plants [*foliage*] in the garden were burnt". This was because of the harmattan season when fallen dry foliages were at extremely low humidity, and thus intensified the spread of the fire, and negatively affected the ecosystem and ecosystem services of the Garden. However, the ecological loss was regained through natural precipitation of the following rainy season. This shows the great impact of climatic factors and dynamics on urban ecological diversities and their ecosystem services. Except for the evergreen trees, most herbaceous plants dry up during dry seasons and harmattan periods which reduce the green quality of the Garden, but reverts back during the rainy season. This type of ecosystem disservice is often neglected in the literature on urban ecosystem disservice. Davoren and Shackleton (2021) decry a general lack of studies that take enough cognizance of ecosystem disservices in the Global South. They propose the need for a framework that simultaneously incorporates ecosystem services and disservices in the context of the Global South to enhance formulation of adequate policies for their management in urban planning and goveranance.

2.6.3 *University of Lagos Lagoon Front Resort/Park, Lagos, Nigeria*

Ecological urbanism of Yoruba cities is imbued with all forms of landscapes and waterscapes with attendant multiple ecosystem services. The reason is that these cities were founded based on different landscape characters which were city-generating factors. Prominent among these are non-man-made, including topographical, hydrological, geological, ecological, floral, and fauna landscapes. The "man-made city-generating factors include all means of transportation, economic and sociocultural opportunities"(Adedeji and Fadamiro 2018, p.798). Hills, rocks, slopes, and valleys are common topographical landscapes that enhance ecological urbanism. Streams, rivers, lakes, lagoons, oceans, and seas are central to the formation of some other cities. All these ecological components of Yoruba cities are significant in defining placeness in different climes. In the case of metropolitan Lagos, the importance of lagoon ecosystem services cannot be overemphasized. Even the megacity derived its name from its environments being characterized by lakes or lagoons. At the interface of land and ocean, Lagos is exemplary as a major coastal and ecological Yoruba city among the coastal network of cities along the Atlantic Coast of West Africa. According to the World Population Review (WPR 2021), the human population of metropolitan Lagos alone is 14,862,11 which exclude the non-metropolitan parts of Lagos State. In this subsection, University of Lagos Lagoon Front Resort/Park is discussed as a contact point to make connections to the ecosystem services of Lagos Lagoon complex. In particular, how the Lagos Lagoon has created a unique place identity for the entire Lagos and campus of the University of Lagos as epicenters of ecological uniqueness are meta-narrated.

The network of lagoons in Lagos urban space consists of nine lagoons namely "Yewa, Ologe, Badagry, Iyagbe, Lagos, Kuramo, Epe, Lekki, and Mushin" (Uduma-Olugu and Oduwaye 2010, p. 757) with Lagos Lagoon being the biggest and directly connected to the Atlantic Ocean. This complex lagoon system stretches from the Benin Republic to Nigeria and has been referred to as Western Nigerian lagoons before (Webb 1958; Hill and Webb 1958). The Lagos lagoon waterfront is made up of many communities that include "Makoko, University of Lagos, Ilaje, Oworonsoki, Ogudu, Bayekun, Agboyin, Moba, Ofin, Ikorodu, Ibeche, Aja, Lekki peninsula, Banana Island and Ikoyi" (Uduma-Olugu and Oduwaye 2010, p. 757). The Lagos and Lekki lagoons are flanked by ecotone spaces of mangroves and edged by swamp forests. These are unique meeting points of aquatic and terrestrial ecosystems which are very rich in urban biodiversities. Uduma-Olugu and Oduwaye (2010) posit that the Lagos lagoon is "a water body subject to tidal waves" and "it is directly linked to the Atlantic Ocean on the south and technically ends around the Palaver Islands on the east, while the lagoon continues to the Epe lagoon which eventually opens out to the ocean again, further along the Lagos coast" (p. 757). The Lagos lagoon is a large water body with surface area of 6354.798 square kilometers at the heart of Lagos metropolis and it is made up of Lagos harbour, metropolitan, and Epe division segments (Oyedele and Momoh 2009). The lagoon enjoys the inflow of fresh water from Ogun River and Osun River through the north and east corridors which set its ecotone parameters like salinity thereby making it an habitable aquatic ecosystem for benthic invertebrates (Fajemila et al. 2020). At some points, the lagoon hosts white mangrove which is unique only to the lagoon environment and can not be found elsewhere since it does not grow in fresh water. Kjerfve (1994) submits that lagoons in coastal locations are greatly productive blackish-watersystems that are valuable for their wide range of ecosystem services economically and ecologically. These urban aquatic ecosystems also offer multiple ecosystem services that support the livelihood and wellbeing of city dwellers (Velasco et al. 2018).

Lagoons are estuaries and natural water bodies positioned between land and ocean backing each other, where there is a mixture of fresh water from inflows and salt water of the ocean. They are coastal natural environments where harbour, jetty, or wharf can be created artificially for loading maritime services. Lagoons could be close, open, or semi open. Lagos lagoon is open to the Atlantic Ocean with six-hourly changes between low and high tides covering spaces depending on the gradient of the shore. Therefore, water from the Atlantic Ocean moves to Lagos lagoon during high surges and recedes during low tides (Oludayo et al. 2021). Ebute Metta, literally meaning three harbours, is located on the lagoon from the colonial era as a docking site for landing of cargo at Iddo, Otto, and Oyingbo. It has colonial complicity as the landing point of the British and Portuguese where in-direct rules through the traditional kingship system was first established, and therefore a historical space in the socio-cultural ecosystem services of Lagos lagoon (Badejo et al. 2014). Ebute Metta is also important as the railway terminal from Lagos to northern parts of Nigeria, making it a crucial socio-economic service hub of the lagoon area (Oni and Okanlawon 2012). The ecosystem services of maritime activities at Ebute Metta area of Lagos lagoon make it central to the maritime economy of Nigeria as a great boost to the economic resource and foreign earning power of the country. A marine ecologist at the University of Lagos explained during an interview with him on the campus of the University that Lagos lagoon is key to shipping trade, cargo goods and services in Nigeria. "That is what the Lagos lagoon provides for us", he added. This is because of the presence of Apapa Wharf in the Lagos lagoon which is the largest wharf in West Africa. The direct connection of Lagos lagoon itself to the Atlantic Ocean unlike all the other lagoons makes it the most economic, most central, and defines its uniqueness for maritime ecosystem services. Being the lowest pond in the whole of Yoruba nation cities of Southwest Nigeria makes Lagos lagoon the recipient of all waters from all the rivers in the region through gravitational force.

The University of Lagos is the only University in Yoruba cities and entire Nigeria with a coastal campus situated by a large body of water like the Lagos lagoon. Specifically, the Department of Marine and Marine Sciences is fittingly located near the Lagos lagoon within the campus with a scenic view to a host of saline mangrove swamp ecosystems of Lagos lagoon that are very rich in biodiversities of flora and fauna, including microbes like bacteria, algae, and fungi. Thatoi et al. (2013) in a review provide a deep insight into the ecological role of microbes in mangrove ecosystem. The authors submit that mangrove microbes are not only significant constituents of mangrove ecosystems, but also "play a very critical role in creating and maintaining this biosphere". This is aside from ecosystem services like being "a source of biotechnologically valuable and important products" (ibid, p. 1). Particularly, they are critical actors in the "various steps of decomposition and mineralization of leaf litter" nutrients recycling, production and consumption of "gases that affect global climate", and destruction of pollutants. Others are treatment of "anthropogenic wastes" and "biological control of plant and animal pests" (ibid, p. 1). Through industrial biotechnological ecosystem services, the mangrove microbes are:

> a major source of antimicrobial agents and also produce a wide range of important medicinal compounds, including enzymes, antitumor agents, insecticides, vitamins, immunosuppressants, and immune modulators (Thatoi et al. 2013, p. 1).

The scenic ecological view of the Lagos lagoon from the Lagoon Front Park is enhanced by the presence of the Third Mainland Bridge which is the longest bridge in Africa being 11.8 km (Adekunle et al. 2012, p. 136) long until 1996 when the Cairo Bridge was completed. The Third Mainland Bridge is also the longest of the three bridges that connect Lagos Islands with the Lagos Mainlands while Eko and Carter are the two other bridges. Today, it remains the longest bridge in Nigeria. The daily heavy vehicular volume that plies the bridge combined with its waterscape background adds an alluring and picturesque scenic view to the Lagos lagoon from the Park. The view is dynamic through the mechanisms of its dominant coastal waves, land breeze, and sea breeze at different times of the day which produce a wide range of therapeutic soothing somatosensation.

Furthermore, the University of Lagos Lagoon Front Resort/Park is a recreation ground of magnificient ecological milieu and rich ecosystem services in a special campus setting as one of the wonders of this coastal University. The terrestrial-aquatic ecotone transition ecosystem of the Park is evident in the numerous crab holes that dotted the site. The location of a mini zoological garden with monkeys and tortoise (*Testudo kleinmanni*, F. Testudinidae), and botanical garden with many plant species near the Park complete the uniqueness of its ecosystem. The marine ecologist interviewed attests to the ways by which the Park contributes greatly to the campus life of the students and entire University community. "You see them reading there sometimes, doing one or two fun activities there", he affirmed. The resort is a resourceful students' base that is regulated by the Dean of Student Affairs for the authorization of formal group activities. To the marine ecologist, Lagos lagoon is a lovable, magnificient piece of nature that could be maximised for "the benefit of humanity". According to him, this is because of its "wonderful resources, living and non-living" and their attendant "great benefits economically and ecologically". Within these spheres, Lagos lagoon is ecologically suitable as a Marine Protected Area (MPA) for nature conservation, tourism, and "power generation", he added. The lagoon fosters human understanding of the semiosis of life mantras like "time and tide wait for no man" through a natural scenery of the ocean tidal waves as a socio-cultural ecosystem service. As sites of sand mining, Lagos lagoon is important for the provisioning ecosystem service of building sand for the constantly evolving built urban spaces of metropolitan Lagos. This is naturally sustainable through the process of beach nourishment, renourishment, or sand replenishment, if the principles of responsible consumption of nature are engaged. Sand mining regulation in lagoons is in conformity with Goal 12 of the United Nations SDGs on sustainable use of natural resources. For this reason, the marine ecologist emphasized that ecological problems would arise at the lagoon if "more than what nature can replace" is mined.

Moreover, Lagos lagoon offers educational benefits as a form of cultural ecosystem services to the students of Marine and Marine Sciences of the University of Lagos. For instance, diploma, undergraduate, and graduate students use the lagoon as an outdoor laboratory for their practicals on most subjects. These include subjects like coastal ecosystems, tourism, oceanography, metrology, pollution, shipping, cargo management, and marine resources, among many others. This makes the Lagos lagoon very central in providing educational ecosystem services to the University of Lagos community.

Lagos lagoon is important for providing means of water transportation for intra-metropolis of Lagos through commercial boating as a relatively faster means of transportation compared with road transportation in the highly vehicularly-busy and usually clumsy commercial city. Although prohibited, some students access the University of Lagos Lagoon Front Park from other parts of Lagos like Oworonsoki through canoe. The marine ecologist gave a practical personal experience of travelling on water "from Victoria Island to Badagry in one hour thirty minutes" which would normally take "almost three hours" by road. This makes "transport along the whole of Lagos State" easy even "to the Republic of Benin through the water". The same ease applies to Ondo and Ogun States as they are all connected to the Lagos lagoon complex. The transport ecosystem services of the Lagos lagoon also include the water transportation of timber logs. For instance, timber logs are transported by water from Edo State of Nigeria through Lekki lagoon to Lagos lagoon and sold "under the Third Mainland Bridge towards the Ijora area. It is fascinating!", he exclaimed. Barbier (2017) identifies water transportation as important economic ecosystem service as he differentiates between ecosystem goods and ecosystem services. This is not only evident in Lagos lagoon but also central to the economic life of the entire Nigeria through the Lagos Port Complex of the Nigerian Ports Authority. Also, it contributes largely to the description of the entire Lagos State as the commercial center of Nigeria. This Port Complex is not only the largest sea port of Nigeria but the oldest dating back to 1913 when it was established and its first four deep water berths had their constructions commenced in 1921. While the port is connected to intermodal rail, water, and road transport systems, it also has five private terminals. In view of its very high congestion at the oil-boom in 1975, the Tin Can Island Port occupying 73 hectares of seascape was established in 1976, completed, and commissioned on the 14th of October 1977 (Nigerian Ports Authority, NPA 2021). These ecosystem services are additional services to the "fishing in the lagoon". The marine ecologist at the University of Lagos provided some insights into the sustainable fishing methods in the Lagos lagoon that:

> They also do fishing in the lagoon. You can see those things, those structures in the water. They call them fish culture (*Acadja*). They are Fish Aggregating Devices (FADs), so they stay more. It is a form of harbour culture. You don't have to put any feed inside. Natural culture, natural form of harbour culture. You don't have to put any feed, just make a structure that attracts organisms to come and stay there. They will think you have provided some shelter and sanctuary, so they come, especially the small ones that are avoiding the big ones, so they come and stay there saying that they have seen two bedroom flat and three bedroom flat, they stay for a while. After you come and take them but not all of them will come. It is a sustainable way of fishing but if you go there and you take mosquito net, you have pack everything. You have carry [sic] father, mother, children, uncle, youngster, teenager, and everything you have taken.

It is for these reasons Lagos lagoon is renown for commercial fishing. The Makoko area of Lagos is a 200 years old fishing village built on the bank of the Lagos lagoon. Like many other areas of Lagos, Isale Eko and Enu Owa are other areas that enjoy fishing as significant provisioning ecosystem services in the coastal city.

With these overall description of the ecosystem services of Lagos lagoon, there is the need to return the narrative to the significance of the lagoon to the University of Lagos Lagoon Front Resort/Park. The lagoon has not only popularized this park, it has made it unique for numerous ecosystem services at the micro and macro levels. For this reason, there are almost one-thousand eight-hundred google reviews of the Park as of the point of writing. These reviews are still in addition to the numerous appraisals of the Park by visitors and tourists in many media including published dailies. For instance, Ekechukwu (2018) presents a summary of the significance of the ecosystem services of this Park as follow:

> The University of Lagos waterfront is a major tourist attraction to the university community. Its tranquil setting explores natural ambience. The inspiration that flows from the place is wistful. On a usual day, a good number of people visit it for various reasons. Being one of the most popular leisure spots at the campus, students come there to read, some on romantic picnics, while others for adventures and licentious trysts. Yet some traverse there just to meditate on their academic misfortunes and woes.

This description is not only apt, it reveals most of the onsite observations of the Park. The entire landscapes of the University campus constituted part of a study of the campuses of the six Federal Universities in Yoruba cities (Adedeji 2018). The study conceptualized the University campuses as cities in their own rights. It assessed their parks and gardens (zoological, botanical), natural spaces like conservation forests/woods, water edges (water fronts), and sports fields (sports pitch, playground). Others are soft landscapes like plant hedges, greenways, shade trees, lawns, shrubs, oil palms, and ornamental palms. The University of Lagos Lagoon Front Park/Result was rated second best recreation site only after the foremost and oldest University of Ibadan Botanical Gardens as campus recreation green landscapes (Adedeji 2018; Adedeji et al. 2020). The ambience of outdoor and indoor environments in the settings of the Park can only be described as soothing coupled with the most fascinating scenic views of nature. In terms of these multifarious ecosystem services enhanced by the micro-climatic regulation, the Lagoon Front Park is a perfect meeting point of green and blue ecosystems in nature. The combination of green spaces of evergreen shading trees and the water body of the lagoon is a perfect match for desirable outdoor recreation for all categories of visitors. In addition to the use of the park by the University community, senior citizens from the Old People's Home in Lagos explore it to enjoy natural feel. Such a nuanced experience of nature has been described in different ways by different visitors to the park in their google reviews of it (Ekechukwu 2018).

One of the visitors, as reported by Ekechukwu (2018) was named Prosper. He visited the park five times within a semester. He was attracted to the park by its natural setting of trees which makes it a cool relaxation space and revealed that the site serves as centre for students to bury their academic miseries and worries. He cited his own case of academic challenge of failing a class test and had to be at the park at that fifth visit to cool off his heated academic displeasure. Little wonder that the park is very popular for the patronage of all categories of social groups for appreciation of nature including kids, young, and aged people within the University campus, its entire Akoka neighbourhood, and far beyond. The Lagoon Front Park is a major attraction of people to the campus and first-timers are naturally eager to be there being also a secure and safe space. These account for the continual presence of people at the brink of the lagoon within the park all the time for different recreational activities in view of its tranquility and serene setting. Visitors to the Park were often seen in solitudes, pairs, and groups of three or more. It is a greatmemorial site for many visitors who also described it as a wonderful place for nature appreciation. Alex grew up in the campus of the University of Lagos and recalled his childhood experiences of the Park and the Lagoon Front. He described it as a place for provisioning ecosystem services of crabs which have numerous super-families, families, and species. He also emphasized how the lagoon front at his adult stage was important as a serene environment that could support any intellectual engagements and overall life productivity. In this way, the Lagoon Front serves as a childhood memorial landscape for him. This is an important socio-cultural ecosystem service towards cognitive development and formation of worldviews about nature. Such recollection of childhood connection with nature environments has been discovered to play significant role in forming nature conservation activism, environmental preferences, and general environmental behaviour and perception in adulthood. Jensen and Olsen (2019) advance that "childhood nature experiences are correlated with pro-environmental preferences in the adult population" (p. 48). Eastwood et al. (2021) also found a close association between human-nature interaction in childhood and adolescence and pro-environmental behaviour that enhances biodiversity conservation and its loss. In fact, ecological worldviews and their nature identities are shaped during childhood play and are critical in framing behaviour on nature conservation and pro-environmental behaviour in later life (Kunchamboo et al. 2021). These make play in nature spaces during childhood and early life developmental stages and ecological-psychological identities in adulthood to be strongly correlated (McGlinn 2019). The two have been linked with how emerging adulthood develops environmental activism (Matsuba et al. 2017).

Furthermore, many visitors to the University of Lagos Lagoon Front Park cherish the sensation of the twilight breeze while relishing the views of the ocean for their picnics, video and photograph shootings including pre-wedding and birthdays, and personal reflections. This makes the Lagoon Front a popular site for families, couples, singles, artists, and social groups who enjoy its quietness aside from the bird watching, canoe paddling, and fishing along the lagoon. The Park offers magnificient, splendous, astounding, aesthetically-pleasing, and natural space for all varieties of activities. These include hanging out, gisting, observing nature and wildlife in their natural habitats (like squirrels, crabs and mudskippers), and getting inspiration. Other socio-cultural ecosystem services of the Park include meditation, carrying out religious personal quiet time and devotion, or meeting other people for all purposes. It is a prayer site for individuals and religious campus fellowship groups who make its ambient material nature as a point of connection to the divine and supernatural nature of the heavenly. As an eco-psychological site, it facilitates quick and speedy delivery of intellectual assignments, goals, and pursuits. This makes it suitable for the patronage of studious students for individual readings, group academic discussions, and last minutes revision for imminent examinations and class tests as revealed by some visitors. Igiri was a visitor to the Park as a song writer and attested to how the beauty, calmness, and serenity of the Park enhanced his productivity in composing many songs with his musical instrument in the Park. Another visitor recalled how the airy, serene, and natural qualities of the Park made it a relaxation and meditation site for overcoming stress by spending most undergraduate time there. The Lagoon Front Park is also popular as a tourist destination for primary and secondary school students on excursions and out-of-class learning expeditions. The locational advantage of the Lagoon Front on the Lagos Mainland makes the Park well-suited for its potentials as a major tourist attraction and relaxation spot on the Mainland. These can even include boat expedition in the perception of Okpah, another visitor. For Olawale, the Park is a charming place to enjoy nature while Miriam portrayed it as comfy and cool, being generally referred to as live garden by most students of the University of Lagos. Perceptual descriptions of the Park by visitors include: superb view of the lagoon, an aquatic environment, sublime nature at its zenith, lovely place to make new friends, and connection space with nature for inner peace. Others are nature conservation space with a flavour of nature, a rewarding reading garden, soothing milieu to the soul, plants and trees space with innovatively arranged seating, and indeed a pristine nature zone with oozing refreshing breeze and fresh air from the lagoon. These experiences of the visitors to the Park are infused with its conviviality in terms of the availability of nibbles, bottled beverage drinks, and ice creams as light refreshments made available for purchase onsite.

Despite all these ecosystem services of Lagos lagoon, its major ecosystem disservice is the social malaise of reported cases of suicide by some Lagos dwellers. These victims of ecosystem disservices deliberately terminated their lives by jumping into the lagoon from the Third Mainland Bridge. Pages of Nigerian newspapers (e.g. Vanguard, The Sun, The Guardian) are replete with such suicide and attempted suicide cases that have attracted negative appellations to the Lagos lagoon as the "lagoon of death" (Usman 2020), and the "Lagos dying spot" (Enyoghasu 2019). Many of the suicidal attempts and events were revealed to be shrouded in mysteries (Odita 2017; Akoni 2019). The interviewed marine ecologist equally emphasized that the suicidal events at the Lagos lagoon might as well suggest that "there are devils there" which might be responsible for those that "are going [*there*] to commit suicide". His assertion is situated within the notion of mystic urban nature, but as mystic ecosystem disservice. Another very crucial ecosystem disservice of Lagos lagoon is that it contributes greatly to the greenhouse gases through the huge maritime operations of water transportation at the Lagos Port Complex. Olukanni and Esu (2018) estimated the "greenhouse gas emissions from portvessel operations at the Lagos and Tin Can ports" in the first two quarters of the year 2017. They discovered that "approximately 16,335 t [*tones*] and 773 t [*tones*] of CO_2 [*Carbondioxide*] were produced and emitted during anchorage and while passing lock gates movements, respectively", just as "644 t [*tones*] was emitted through maneuvering to the dock movement" (p. 1). These are inevitable ecosystem disservices at the ports as cargos are moved from the ocean coast about 400 kilometers away in the process of which Carbondioxide is emitted into the air as "a major ozone depleting substance discharged by ships with respect to amount and global temperature alteration potential" (ibid, p. 2). The University of Lagos Lagoon Front Park also has ecosystem disservice as a site of anti-social activities like substance smoking and nocturnal activities of gangs especially towards the brink of the lagoon at odd hours. Insect bite, like mosquitoes and ants, and harmless but freely roaming crabs constitute ecosystem disservices to some visitors who have phobia for them. Also, in view of the inland water flow into the lagoon along with solid wastes like empty plastic bottles, the lagoon contributes to the dirtiness of the ocean shores as these materials litter it through ocean waves.

Consequent upon the Anthropocene, continual yearning for access to the urban ecosystem services offered by the ecological urban spaces considered in this chapter calls for policy formulation on ecosystem services in the region. This could be institutionalized through the Federal (and States) Ministry of Tourism, exploring the opportunities of public-private initiatives towards the maintenance and conservation of the spaces for future generations. The next chapter of the book discusses the diverse ways Yoruba people relate with Green Environments to access their ecosystem services for wellbeing and the mechanisms that frame these services.

References

Adebanwi W (2008) Death, national memory and the social construction of heroism. J Afr Hist 49(3):419–444

Adebayo M (2019) UI short of 1, 834 staff, lecturers overstretched. Daily Post, May 25. Accessed 04 Oct 2021 from https://dailypost.ng/2019/05/25/ui-short-1834-staff-lecturers-overstretched-vc-cries/

Adedeji JA (2018) Assessment of open spaces landscape in federal universities in south-west, Nigeria. PhD Thesis, department of architecture, school of post graduate studies, Federal University of Technology, Akure

Adedeji JA (2020) Urban biocultural identity of Yorubas: intersection of philosophy and nature for wellbeing at Osun sacred Grove UNESCO site, Osogbo, Nigeria. In: Cocks M, Shackleton C (eds) Urban nature: enriching belonging, wellbeing and bioculture, 1st edn. Routledge, London, pp 51–66

Adedeji JA, Adeboyejo TA (2011) Environmental hazards, disasters and pollution: global, African and Nigerian patterns. LAMBERT Academic Publishing Gmb H & Co. KG, Saarbrücken. ISBN: 978-3-8454-7587-5

Adedeji JA, Fadamiro JA (2018) Urbanisation forces on the landscapes and the changing value-systems of Osun sacred Grove UNESCO site, Osogbo, Nigeria. *Landsc Res* 43(6):798–816

Adedeji JA, Fadamiro JA, Odeyale TO (2020) Design toolkits for campus open spaces from post-occupancy evaluations of federal universities in south-West Nigeria. Built Environ Proj Asset Manag 10(2):296–311

Adekunle AA, Adewumi JK, Bello NJ, Akinyemi JO (2012) Stormwater characteristics on 3rd mainland bridge Lagos, Nigeria. J Appl Technol Environ Sanit 2(3):135–144

Aerts R, Honnay O, Van Nieuwenhuyse A (2018) Biodiversity and human health: mechanisms and evidence of the positive health effects of diversity in nature and green spaces. Br Med Bull 127(1):5–22

Ahmad Sayuti N, Zamri Azizan M, Fazlan Ahmad Zamri A, Rizaimy Shaharudin M, Abdulwahhab Khalaf O (2019) The development of typologies, design patterns and scaled models bio-inspired by mangrove species. Int J Adv Sci Technol 28(13):451–464

Akoni O (2019) Middle aged man rescued from committing suicide on third mainland bridge. Vanguard, October 24, 2019. Online. Accessed on 28 Oct 2021 at https://www.vanguardngr.com/2019/10/middle-aged-man-rescue-from-committing-suicide-on-third-mainland-bridge/

Aweto AO (2001) Trees in shifting and continuous cultivation farms in Ibadan area, southwestern Nigeria. Landsc Urban Plan 53(1–4):163–171

Badejo OT, Olaleye JB, Lademomi AS (2014) Tidal characteristics and sounding datum variation in Lagos state. Int J Innov Res Stud 3(1–23):2319–9725

Barbier EB (2017) Marine ecosystem services. Curr Biol 27(11):R507–R510

Barker A, Crawford A, Booth N, Churchill D (2019) Everyday encounters with difference in urban parks: forging 'openness to otherness' in segmenting cities. Int J Law Context 15(4):495–514

Burley BA (2018) Green infrastructure and violence: do new street trees mitigate violent crime? Health Place 54:43–49

Callaghan A, McCombe G, Harrold A, McMeel C, Mills G, Moore-Cherry N, Cullen W (2021) The impact of green spaces on mental health in urban settings: a scoping review. J Ment Health 30(2):179–193

Campos P (2010) The concept of "educational campus" and its application in Spanish Universities. CELE Exchange, 2010/8: OECD Publishing

Carmona M, Heath T, Oc T, Tiesdell S (2003) Public places-urban spaces: the dimensions of urban design. Architectural Press, Oxford

Chen Y, Xu Z, Byrne J, Xu T, Wang S, Wu J (2021) Can smaller parks limit green gentrification? Insights from Hangzhou, China. *Urban For Urban Green* 59:127009

Crompton JL (2005) The impact of parks on property values: empirical evidence from the past two decades in the United States. Manag Leis 10(4):203–218

Crompton JL, Nicholls S (2019) The impact of park views on property values. Leis Sci:1–13

d'Acci L (2019) Quality of urban area, distance from city Centre, and housing value. Case study on real estate values in Turin. Cities 91:71–92

Davoren E, Shackleton CM (2021) Urban ecosystem disservices in the global south. In: Urban ecology in the global south. Springer, Cham, pp 265–292

de Kleyn L, Mumaw L, Corney H (2020) From green spaces to vital places: connection and expression in urban greening. Aust Geogr 51(2):205–219

Dominey-Howes D (2018) Hazards and disasters in the Anthropocene: some critical reflections for the future. Geosci Lett 5(1):1–15. https://doi.org/10.1186/s40562-018-0107-x

Dow JM (2022) Nature aesthetics. Philos Compass 17(5):e12829

Eastwood A, Juárez-Bourke A, Herrett S, Hague A (2021) Connecting young people with greenspaces: the case for participatory video. People Nat 2021(00):1–11. https://doi.org/10.1002/pan3.10236

Ekechukwu F (2018) Unilag waterfront: an untapped tourist desination. This day, October 13, 2018 online. Accessed 29 Oct 2021 at https://www.thisdaylive.com/index.php/2018/10/13/unilag-waterfront-an-untapped-tourist-destination/
Eliasson I, Knez I, Fredholm S (2018) Heritage planning in practice and the role of cultural ecosystem services. Herit Soc 11(1):44–69
Enyoghasu L (2019) Mystery of Lagos dying spot. *The sun*, 9th February 2019. Sunnewsonline. Accessed on 28 Oct 2021 at https://www.sunnewsonline.com/mystery-of-lagos-dying-spot/
Fajemila OT, Sariaslan N, Langer MR (2020) Spatial distribution of benthic foraminifera in the Lagos lagoon (Nigeria): tracing the impact of environmental perturbations. PLoS One 15(12):e0243481. https://doi.org/10.1371/journal.pone.0243481
Fasihi H, Parizadi T (2020) Analysis of spatial equity and access to urban parks in Ilam, Iran. J Environ Manag 260:110122
Fisher JC, Irvine KN, Bicknell JE, Hayes WM, Fernandes D, Mistry J, Davies ZG (2021) Perceived biodiversity, sound, naturalness and safety enhance the restorative quality and wellbeing benefits of green and blue space in a neotropical city. Sci Total Environ 755:143095
Fletcher R (2015) Nature is a nice place to save but I wouldn't want to live there: environmental education and the ecotourist gaze. Environ Educ Res 21(3):338–350. https://doi.org/10.1080/13504622.2014.993930
Gilboa S, Jaffe ED, Vianelli D, Pastore A, Herstein R (2015) A summated rating scale for measuring city image. Cities 44:50–59
Gormley MJ (2021) The end of the Anthropocene: ecocriticism, the universal ecosystem, and the Astropocene. Rowman & Littlefield
Hamon RB (2018) Garden and "wilderness": an ecocritical exploration of gen. 2: 4b–3: 24. Bible Crit Theory 14(1):63–86
Harrison EA (2019) Tree species of Lekki conservation center, Lagos state, Nigeria. Int J Biosci 15:308–314
Hill MB, Webb JE (1958) The ecology of Lagos lagoon II. The topography and physical features of Lagos harbour and Lagos lagoon. Philos Trans R Soc Lond Ser B Biol Sci 241(683):319–333
Hinchliffe S, Whatmore S (2006) Living cities: towards a politics of conviviality. Sci Cult 15(2):123–138
Hosaka T, Numata S, Sugimoto K (2018) Research note: relationship between childhood nature play and adulthood participation in nature-based recreation among urban residents in Tokyo area. Landsc Urban Plan 180:1–4
Howard E (2010) To-morrow: A peaceful path to real reform. Cambridge University Press
Hurly J, Walker GJ (2019) Nature in our lives: examining the human need for nature relatedness as a basic psychological need. J Leis Res 50(4):290–310
Jamali AA, Tabatabaee R, Randhir TO (2021) Ecotourism and socioeconomic strategies for Khansar River watershed of Iran. Environ Dev Sustain 23:17077–17093. https://doi.org/10.1007/s10668-021-01334-y
Jensen AK, Olsen SB (2019) Childhood nature experiences and adulthood environmental preferences. Ecol Econ 156:48–56
Jim CY, Chen WY (2010) External effects of neighbourhood parks and landscape elements on high-rise residential value. Land Use Policy 27(2):662–670
Karst HE, Nepal SK (2022) Social-ecological wellbeing of communities engaged in ecotourism: perspectives from Sakteng wildlife sanctuary, Bhutan. J Sustain Tour 30(6):1177–1199
Kjerfve B (1994) Coastal lagoon processes. Elsevier Oceanography Series. Elsevier, Amsterdam
Knowlton JL, Halvorsen KE, Flaspohler DJ, Webster CR, Abrams J, Almeida SM, Arriaga-Weiss SL, Barnett B, Cardoso MR, Cerqueira PV et al (2021) Birds and bioenergy within the Americas: across-national, social–ecological study of ecosystem service tradeoffs. Land 10:258. https://doi.org/10.3390/land10030258
Koole SL, Van den Berg AE (2005) Lost in the wilderness: terror management, action orientation, and nature evaluation. J Pers Soc Psychol 88(6):1014–1028

Kunchamboo V, Lee CKC, Brace-Govan J (2021) Cultivating nature identity and ecological worldviews: A pathway to alter the prevailing dominant social paradigm. J Macromark 41(3):484–505
Leal Filho W, Tripathi SK, Andrade Guerra JBSOD, Giné-Garriga R, Orlovic Lovren V, Willats J (2019) Using the sustainable development goals towards a better understanding of sustainability challenges. Int J Sustain Dev World Ecol 26(2):179–190
Machnik A (2021) Ecotourism as a Core of sustainability in tourism. In: Lubowiecki-Vikuk A, de Sousa BMB, Đerčan BM, Leal FW (eds) Handbook of sustainable development and leisure services, World sustainability series. Springer, Cham. https://doi.org/10.1007/978-3-030-59820-4_15
Maréchal L, MacLarnon A, Majolo B, Semple S (2016) Primates' behavioural responses to tourists: evidence for a trade-off between potential risks and benefits. Sci Rep 6(1):1–11
Matsuba MK, Alisat S, Pratt MW (2017) Environmental activism in emerging adulthood. In: Paddila-Walker LM, Nelson LJ (eds) Flourishing in emerging adulthood: positive development during the third decade of life. Oxford University Press, Oxford, pp 175–201
McGlinn MT (2019) Childhood nature play and the adult ecopsychological self: a mixed methods study. Michigan School of Professional Psychology
McHarg IL (1997) Ecology and design. In: Thompson GF, Steiner CF (eds) Ecological design and planning. Wiley, New York, pp 321–332
Mettinger TND (2007) The Eden narrative: A literary and religio-historical study of genesis 2–3. Eisenbrauns, Winona Lake, p 2007
Nesticò A, Sica F, Endreny T (2020, May) Real estate values and ecosystem services: correlation levels. In: International symposium: new metropolitan perspectives. Springer, Cham, pp 802–810
Nichols A (2011) Beyond romantic ecocriticism: toward urbanatural roosting. Springer
Nigerian Ports Authority, NPA (2021) Port of Lagos Quays. Retrieved from https://nigerianports.gov.ng/lagos-port/
NPC – National Population Commission (2006)
Nwachukwu CE, Olufunmilayo EO, Otor VO, Yakubu AO, Akingbade AE, Odefemi OF, Ikwunne BN et al (2021) Common mental health problems and associated factors among medical students of University of Ibadan, Nigeria. *J Ment Health* 2021:1–8
Nwokorie EC, Adeniyi EE (2021) Tourists' Perception of Ecotourism Development in Lagos Nigeria: The Case of Lekki Conservation Centre. *Turizam 25*(1):11–30
Odita S (2017) Doctor commits suicide, jumps into Lagos lagoon. The Guardian, 20 March 2017. Online. Accessed on 28 Oct 2021 at https://guardian.ng/news/doctor-commits-suicide-jumps-into-lagos-lagoon/
Oludayo AA, Chidi OV, Kayode OF (2021) Triangulation approach for mapping groundwater suitability zones in coastal areas around Lagos, Nigeria. Using multi-criteria decision-making technique. NRIAG J Astron Geophys 10(1):423–442
Olukanni DO, Esu CO (2018) Estimating greenhousegas emissions from port vessel operations at the Lagos and tin can ports of Nigeria. Cogent Eng 5(1):1507267. https://doi.org/10.1080/23311916.2018.1507267
Oni SI, Okanlawon KR (2012) An assessment of the usage of the Lagos mass transit trains. Int J Railw 5(1):29–37
Oppermann S (2011) The future of ecocriticism: present currents. In: Oppermann S, Özdağ U, Özkan N, Slovic S (eds) The future of ecocriticism: new horizons. Cambridge Scholars Publishing, Newcastle upon Tyne, pp 14–29
Oyedele KF, Momoh EI (2009) Evaluation of sea waterintrusion in freshwater aquifers in a lagoon coast: a case study of the University of Lagos Lagoon, Akoka, Nigeria. N Y Sci J 2:32–42
Palliwoda J, Kowarik I, von der Lippe M (2017) Human-biodiversity interactions in urban parks: the species level matters. Landsc Urban Plan 157:394–406
Parsons G (2007) The aesthetics of nature. Philos Compass 2(3):358–372

Pickett ST, Cadenasso ML, Grove JM (2004) Resilient cities: meaning, models, and metaphor for integrating the ecological, socio-economic, and planning realms. Landsc Urban Plan 69(4):369–384
Ramírez F, Santana J (2019) Philosophy of ecotourism. In: Environmental education and ecotourism, Springer, Cham, pp 27–31
Raven JA, Karley AJ (2006) Carbon sequestration: photosynthesis and subsequent processes. Curr Biol 16(5):R165–R167
Rival L (2021) Trees, from symbols of life and regeneration to political artefacts. In The social life of trees, Routledge, pp 1–36
Rodriguez MB, Simon M (2015) Conceptualizing conviviality in urban landscapes. Athens J Archit 1(4):311–326
Santos MN, Delabie JH, Queiroz JM (2019) Biodiversity conservation in urban parks: a study of ground-dwelling ants (hymenoptera: Formicidae) in Rio de Janeiro City. Urban Ecosyst 22(5):927–942
Saunders ME, Luck GW (2016) Limitations of the ecosystem services versus disservices dichotomy. Conserv Biol 30(6):1363–1365
Soga M, Gaston KJ (2022) Towards a unified understanding of human–nature interactions. Nat Sustain 5(5):374–383
Soga M, Gaston KJ (2020) The ecology of human–nature interactions. Proc R Soc B 287(1918):20191882
Solomon S, Greenberg J, Pyszczynski T (2004) The cultural animal: twenty years of terror management theory and research. In: Greenberg J, Koole SL, Pyszczynski T (eds) Handbook of experimental existential psychology. Guilford Press, New York, pp 13–34
Sonti NF, Campbell LK, Svendsen ES, Johnson ML, Auyeung DN (2020) Fear and fascination: use and perceptions of new York City's forests, wetlands, and landscaped park areas. Urban For Urban Green 49:126601
Spirn AW (2002) The authority of nature: conflict, confusion, and renewal. In: Johnson BR, Hill K (eds) *Ecology and design: frameworks for learning*. Island Press, Washington, DC, pp 29–49
Steiner F (2014) Landscape ecological urbanism'. Sustainable Urban Development Reader, 190
Thatoi H, Behera BC, Mishra RR, Dutta SK (2013) Biodiversity and biotechnological potential of microorganisms from mangrove ecosystems: a review. Ann Microbiol 63:1–19. https://doi.org/10.1007/s13213-012-0442-7
Tibesigwa B, Ntuli H, Lokina R (2020) Valuing recreational ecosystem services in developing cities: the case of urban parks in Dar es Salaam, Tanzania. *Cities* 106:102853
Till KE (2005) The new Berlin: memory, politics, place. University of Minnesota Press, Minneapolis
Uduma-Olugu N, Oduwaye L (2010) The regeneration of Lagos lagoon waterfronts for recreation and tourism. In M Schrenk, VV Popovich, P Zeile, (eds) Real Corp 2010: cities for everyone. Liveable, healthy, prosperous. Proceedings/Tagungsb and Vienna, 18-20 May 2010 (page 757–762) – http://www.corp.atEditors:
Usman E (2020) Lagoon of death where military personnel, doctor, others perished. *Vanguard*, February 22, 2020 online. Accessed 28 Oct 2021 at https://www.vanguardngr.com/2020/02/lagoon-of-death-where-military-personnel-doctor-others-perished/
Van Dijk-Wesselius JE, Maas J, Hovinga D, Van Vugt MVDBA, Van den Berg AE (2018) The impact of greening schoolyards on the appreciation, and physical, cognitive and social-emotional Well-being of schoolchildren: A prospective intervention study. Landsc Urban Plan 180:15–26
Velasco AM, Perez-Ruzafa A, Martinez-Paz JM, Marcos C (2018) Ecosystems services and main environmental risks in a coastal lagoon (Mar Menor, Murcia, SE Spain): the public perception. J Nat Conserv 43:180–189
Voicu I, Been V (2008) The effect of community gardens on neighboring property values. Real Estate Econ 36(2):241–283
Wang HJ (2019) Green city branding: perceptions of multiple stakeholders. J Prod Brand Manag 28(3):376–390

Wang X, Liu J, Liang C, Zhao Z, Feng G, Zhang J (2021) Biodiversity dataset of vascular plants and birds in Chinese urban greenspace. Ecol Res 36:755–761

Warrier RR, Priya SM, Kalaiselvi R (2020) Gmelina arborea–an indigenous timber species of India with high medicinal value: A review on its pharmacology, pharmacognosy and phytochemistry. J Ethnopharmacol 113593

Webb JE (1958) The ecology of Lagos lagoon I. the lagoons of the Guinea coast. Philos Trans R Soc Lond Ser B Biol Sci 241(683):307–318

Weiland S, Hickmann T, Lederer M, Marquardt J, Schwindenhammer S (2021) The 2030 agenda for sustainable development: transformative change through the sustainable development goals? Politics Gov 9(1):90–95

Wong FY, Yang L, Yuen JW, Chang KK, Wong FK (2018) Assessing quality of life using WHOQOL-BREF: A cross-sectional study on the association between quality of life and neighborhood environmental satisfaction, and the mediating effect of health-related behaviors. BMC Public Health 18(1):1–14

World Population Review, WPR (2021) Accessed on 13 Aug 2021 from https://worldpopulationreview.com/world-cities/akure-population

Wu C, Ye X, Du Q, Luo P (2017) Spatial effects of accessibility to parks on housing prices in Shenzhen, China. Habitat Int 63:45–54

Zhang S, Liu J, Song C, Chan CS, Pei T, Yu W, Zhang X (2021) Spatial-temporal distribution characteristics and evolution mechanism of urban parks in Beijing, China. Urban For Urban Green 127265:127265

Zhu J, Lu H, Zheng T, Rong Y, Wang C, Zhang W et al (2020) Vitality of urban parks and its influencing factors from the perspective of recreational service supply, demand, and spatial links. Int J Environ Res Public Health 17(5):1615

Zylstra M, Esler K, Knight A, Le Grange L (2019) Integrating multiple perspectives on the human-nature relationship: A reply to Fletcher 2017. J Environ Educ 50(1):1–10

Zylstra MJ, Knight AT, Esler KJ, Le Grange LLL (2014) Connectedness as a core conservation concern: an interdisciplinary review of theory and a call for practice. Springer Sci Rev 2(1–2):119–143. https://doi.org/10.1007/s40362-014-0021-3

Chapter 3
Ecosystem Services of Yoruba Homegardens – Eco-cultural Indigenous Knowledge System for Wellbeing

Abstract Yoruba homes are one of their most unique cultural representations of ecological connection to nature in everyday life. This is unmistakable in the different types of non-ornamental gardens that are usually in contiguity to the home buildings or in their courtyards. Also, planting of trees around home environments are usual practices. In view of the colonial origin of building and planning regulations in Nigeria like many other former colonies and the global realization of the efficacy of nature-based solutions to biogeophysical problems, understanding of these practices can shed light on the ecosystem services of the gardens to underscore Yoruba homes as multispecies settings. The aim of this chapter is to examine the operational mechanisms of wellbeing through indigenous knowledge system (IKS) of urban homegardens and situate them within the wider discourse of Yoruba ecological urbanism. This was carried out by transcending social, spatial, temporal, economic, intellectual, and religious boundaries in the Yoruba cities. Ecosystem services are discovered to be means of culturally institutionalizing kinship, 'living' memories, landscape management, tangible and intangible heritages. Others are environmental sustainability, soil management, ethnobotany, and general ethnobiological and ecological lifestyles. Likewise, the practices of urban homegardening by Yoruba people are embedded with deep IKS of many branches of agriculture and ecology. These include agronomy, soil science, horticulture, food science, agrometrology, agricultural economics, plant pathology, agricultural entomology, agro-forestry, poultry, fishery, snailry, among many others. These are discovered through interrogation of provisioning, supporting, regulating, and cultural ecosystem services of homegardens. The chapter concludes by arguing for the need to incorporate these IKS of homegardening into a review of the current planning and building regulations for human wellbeing. The current regulations have colonial origins and are situated within worldviews of the Global North. The next Chapter of the book discusses the distinctions between biophilic ideology and sacralisation of natural elements in Yoruba urbanism. This is to argue for the differences between the Global North and South on human relationships with the physical world of ecological elements of urban nature.

J. A. Adedeji, *Ecological Urbanism of Yoruba Cities in Nigeria*, Cities and Nature, https://doi.org/10.1007/978-3-031-34688-0_3

Keywords Yoruba indigenous knowledge system · Homegardens · Provisioning ecosystem services · Medicinal plants · Cultural ecosystem services · Human wellbeing

3.1 Indigenous Knowledge System for Human Wellbeing in Yoruba Cities

Landscape evaluation through the approach of ecosystem services is a vital tool in urban landscape planning and design (von Haaren et al. 2019). The Holy Bible provides a great insight into this theme on homegardens as a perfect picture of ecological urbanism. It says in the verse 19 of Chapter 13 of the Gospel according to Saint Luke that "It is like a grain of mustard seed, which a man took, and cast into his garden; and it grew, and waxed a great tree; and the fowls of the air lodged in the branches of it"(James et al. 1979, p. 149). This verse mirrors and captures the key components and actors of a multispecies urban homegarden for the wellbeing of human and nonhuman nature – man/human, flora/tree, fauna/fowls – in the garden ecosystem. The aim of this chapter is to examine the operational mechanisms of wellbeing strategies through ecosystem services of urban homegardens and situate them within the wider discourse of Yoruba ecological urbanism. This is with a view to revisiting (in order to reconnect with) the IKS of traditional practices and integrating them into modern planning frameworks of homegardens for human wellbeing. Ecological urbanism demonstrated through Indigenous Knowledge System, IKS, is a haven for wellbeing among the Yoruba people. The practise of IKS in the relationship of human and nonhuman nature among the Yoruba people appears to be the most enduring tradition that has not been significantly overrun by modernisation and civilisation forces in both sub-urban and urban places. It is also a unique bridge that spans many layers. These include social, spatial, temporal, economic, intellectual, cultural, and religious divides. The pervasiveness of these forms of ecosystem services transcend social stratification of different groups. While the elite groups have lifestyles characterized by affluence and constant access to continuously evolving and revolving modernism, the outright failure or sometimes ineffectiveness of modern means of wellbeing through western medication has forced them to continue engaging the traditional strategies of IKS for wellbeing. In the case of non-elitist majority, economic challengesappear to make them continue with IKS for the same purpose. The reformation of these strategies through modernized packaging has therefore created a scenario where IKS has become dominant in achieving societal wellbeing. In a similar vein, spatio-temporal variables continue to have no effect on the ecological relationship of Yoruba people with the environment. Specifically, the manifestation of IKS to achieve wellbeing remains significant in all urban settings and at all times with little variations only in the modesof operation. At the intellectual level, the wide spectrum of the least to the highest scholarly groups is embedded with IKS for wellbeing as shall be discovered in the sections that follow. IKS has remained a cultural core that defiles religious boundaries from the indigenous religions to Christianity and Islam. Even among various sects, the belief pattern only manifests in modes of application and mechanism. These mechanisms have been dictated by multiple orientations to the notions of Nature that are essentially rooted in ecological belief systems or what Grainger (2022) calls nature-based spiritualities.

Accordingly, the ecosystem services of homegardens in Yoruba cities are hinged on IKS. There is an inexplicably interwoven relationship of Yoruba people with the natural features in their urban environments. Since the home is the commonest milieu of such socio-ecological and cultural ecosystem services, it provides a special avenue for such daily lifestyles. The yearning of Yoruba people for owner-occupied personal abode surpasses mere provision of dwellings for utilitarian purposes but an environment for cultural expressions. Ecologically, this is realised mainly through homegardens, to demonstrate and explore opportunities of IKS for wellbeing.

In coalescing a definitional approach to understanding the concept of IKS from multiple sources, Nwonwu (2021, p. 141) defined it as "the compendium of knowledge systems that facilitate the pursuit of human wants and needs". These are fundamentally established on all facets of human activities and endeavours, including worldviews, practices, novelties, and know-hows of societies, collections, and people of specific geographical space. They are connected to the "belief systems, life perceptions, and socio-cultural practices" of "indigenous, traditional", and native people "endowed with sophisticated understanding of their culture, ecosystems, language, and local surroundings" which they apply in their daily activities and ways of life (Ibid, p. 141). The innovativeness and inimitability of talents and long-time histories of practices by these groups of people in their very "close interaction with the natural environment across cultures and geographical spaces" make IKS totally indispensable even in this modern time (Seko et al. 2021, p. 185; Lwonga et al. 2011). IKS has also been described to have high potential for the overall development, including socio-environmental and ecological developments, of African people generally and for dealing with globalization challenges (Tondi 2019). An essential part of human need is wellbeing and the means of satisfying ithas been embedded in IKS from time immemorial.

The depth of strategies for wellbeing embedded in IKS of the Yoruba people challenges the rationale for its rejection and ultimate exclusion from mainstream science based on lack of procedural means of describing/analyzing their modes of operation. In contradiction to this claim of science, however, finding solutions to the human-environment problems on sustainability would benefit from multiple methods of inquiry. Of these, the cultural domain is not only very strategic, symbolic, significant, and systematic, it is the embodiment of the operational environment for all others. It is epistemically astounding to note that the value-system of the socio-cultural ecosystem services of the Yoruba people are abundantly corroborated by pharmacognostic expertise with Western perspective. Also, it is very important to note that the medicinal plants that the Yoruba people have identified for generations in their ethnobotanical cultural lifestyles have also been discovered to be essential ingredients in the production of modern Western medications.

The abundant availability of traditionally occurring native in-situ plants despite the continual increase of urbanisation has been enhanced by the importation of exotic ex-situ plants in many urban homes. The love for nature and natural environments serve as impetus for most home owners, especially those that are owner-occupied to devote adequate spaces for developing, growing and maintaining green spaces adjacent to homes. Unlike the western housing setting where green spaces are essentially for ornamental purposes and fauna gardens are kept for only conservation purposes and animal rights to demonstrate love for nature, the green spaces in Yoruba homes serve numerous ecosystem services. These are interwoven with both biophilic and symbiotic, material and mysthical layers, including provisioning, supporting, regulating, and socio-culturalecosystem services.

The ecosystem services of homegardens and gardening practices reported in this chapter showcase the IKS of Yoruba people as being laden with environmental stewardship and rewards in form of derived services. The Yoruba eco-theological value-system is summed up in their belief that "man is a tenant on God's earth" (Idowu 1978, p. 206) which plays out in their care for the environment with the yielding of ecosystem services in return. These environmental endeavours are means of traditionally institutionalizing kinship, 'living' memories, landscape management, cultural heritage, environmental sustainability, soil management, ethnobotany, and general ethnobiological and ecological lifestyles. It is amazing to note that these practices of IKS byYoruba people are embedded with deep traditional understanding of almost all branches of agriculture and ecology including agronomy, soil science, horticulture, agricultural economics and extension. Others are food science, agrometrology, agricultural economics, plant pathology, agricultural entomology, agro-forestry, poultry, fishery, snailry, among many others. These shall be discussed along with the ecosystem services.

3.2 Provisioning Ecosystem Services of Yoruba Homegardens

Home greens in the Yoruba homeland occur mostly as gardens. These gardens are largely non-ornamental. They are means of providing herbal and non-herbal materials for the non-syncretic citizens and metaphysical materials for the syncretic practitioners. Mrs. Faremi O. S., a Yoruba, NCE holder, and retired Head-teacher who hails from Okitipupa had attained the age of 60 years when interviewed in her homegarden at Agbede Street, GRA, Owo. She declared that her knowledge of herbal and gastronomic significances of bitter leaves was ancestral, which she inherited from her mother as a nonmaterial heritage. She equally emphasized the timely and regular fruits that her homegarden providesas essential dietary components for her wellbeing and healthy living. In comparing the homegarden with public markets as sources of vegetables, she pointed out that "during the dry season of the year, when other common vegetables are not available even at the market, this Jerusalem leaf [*plant*] will still be fresh and green, and it will still contain high quantity of chlorophyll". Kaszás et al. (2020, p. 2) emphasized the ecological significance of Jerusalem artichoke (JA) leaf plant (*Helianthus tuberosus L.*) and recalled from multiple scientific sources that the "green leaves of JA are a source of several phytochemicals". According to them, "It is tolerant to many biotic and abiotic stresses" and therefore "widely adapted to diverse and often marginal environments with harsh growth conditions". Wang et al. (2022) equally report that JA is used in indigenous medicine for the treatment of diabetes and rheumatism.

The maternal narrative of Mrs. Faremi O. S. on homegarden idea as a cultural inheritance is similar to the parental story of Madam Esther, a married Yoruba-Itsekiri, and a holder of Bachelor's degree in education. She was at an age of 51 years when interviewed in her homegarden at Warri. To argue for the provisioning sustenance value of the garden, she emphasized the following:

> I grew up with my parents, and my parents, they were into keeping of homegarden, so I learnt it from them. While with my parents, we had sixty to seventy percent of things we needed to use in the house, we were getting them from our homegarden. So we buy very little things that we don't have in the garden, so I got the idea from my parents, so it has become a part of me.

According to her narrative, homegardens are great sources of relief from financial burdens of need to purchase food items for the household. Through this provisioning ecosystem service, Yoruba people are able to cope with meager financial resources, achieve economic sustainability, and still provide satisfactorily for the food requirements of family members. Mrs. Iwapele C. P., a Yoruba school teacher at the age of 44 years was interviewed in her homegarden at Oyo. She also bared her mind on an important motivational factor for establishing the homegarden for ease of access to medicinal herbs and other plant products to be at the disposal of family members. She recalled that the homegarden provided reconnection to the rural lifestyle and upbringing of the husband, though they are now living in the city. In the rural family background of the husband, "they like eating much fruit because they know it helps our body system to function well". Environmental affordances and preferences for nearby nature have proved to be significant contributors to human wellbeing through urban homegardens and green spaces (Clayton 2007; Hadavi et al. 2015). Reflective transfer of childhood environmental preferences in the countryside to an urban adult lifestyle appears to playout in the scenario painted by Mrs. Jeminiwa. Reproduction of rural lifestyle in cities is thus a pathway to developing ecological cityness.

These splendid connections and assemblages of timeframes are valid evidences of socioecological temporalities on which Yoruba urbanism is built. The sociocultural aspiration of Yoruba people to own personal abodes containing homegardens has trajectories whose roots are embedded in their past infancies when their ideologies for 'being' are constructed on complex layers of environmental consciousness. This entrenched mindset portrays the notion of time dynamics of cultural spacethat Song et al. (2008) referred to as "sedimentation theory of time in space" in which he further commented:

> the dynamics of change in a cultural space occurs in the co-present, a place where the reconstructed past is linked with the co-present. It is [*in*] this co-present space that the social construction of cultural space takes place. Some events are retained and defined as belonging to the past and are designed as the old present; other events are modified, refined, or reconstructed in the present and function as the new-present (p. 23).

This notion portrays homegarden in Yoruba urbanism as cultural space. The cultural conception of homegardens also has place attachment and geographical layers as shown in the analysis of Madam Esther who recalled as follows:

> No, I had one previously, although on a smaller scale while I was a tenant before I had my own compound and it has been my longing that when I eventually have my own compound, I was going to do a bigger one, which is why you can see it now.

The idea of homegardening was so strong for her beyond any obstacle. The landlord of her residence at a point in time detested any situation of tenants planting in the compound with the opinion that such tenants can lay ownership claim in the future. This led her to a mode of adaptation by planting *ugwu* and spinach in buckets which proved to be very successful. With this experience, she already developed mental ideas of landscape planning of her future owner-occupied home with garden which became a reality. She pointed out that the homegarden was important to her as much as the home building but with the need to erect the building first before commencing the planting of the homegarden to avoid any need of uprooting.

This perspective on home landscape planning was framed in terms of plot size by Mr. Felix Olusola Olagunju, a married 59 years old Yoruba man from Ekiti with two Master's degrees when interviewed in his homegarden at Ilorin (hereafter simply referred to as Mr. Olagunju). He narrated that the plot and building sizes have direct bearing on homegarden and landscape planning with the emphasis that a minimum of two plots should be ideal or one plot with a small-size home building to allocate enough "space for trees". According to him, the home landscape planning should give priority to the provision of green spaces as well as outdoor sewage installations. Despite the enforcement of planning regulations in the region stipulating a lapsing period of 2 years for approved documents, there is no framework for home green spaces aside from buildable percent areas and airspaces. This mode of procedure mirrors the connection between traditional and modern planning ideologies where formal landscape planning is not enforced like in these scenarios. Coupled with this planning paradigm is the dynamism of green spaces that allow for desirable modifications through re-arrangement, removal, re-planting, and re-establishment of the home landscape. This was illustrated by Mr. Olagunju:

> [...] since I have soursop, almost about ten trees, I'm thinking of reducing the number of pawpaw or soursop [*trees*] to put that apple there to replace it so that with the replacement, as time goes on, I will cut some. I just love innovation and what is medicinal.

The provisioning ecosystem services of homegardens was also a motivational factor derived from the familybackground of Madam Esther. She recalled that "growing up as a child, the homegarden helped my family a lot", her father being a civil servant, "so we got most of the things we needed in the house from the garden without spending money". She continued further that "even buying of meat, we had fowls during festive period or occasionally we get chicken from our small poultry for house use". In view of this, she leveraged on her parental background "to supply what I need in the house through my homegarden, so that has reduced my spending [...] to be able to raise funds to also train my children". With this narrative, the socioeconomic ecosystem services of homegardens are evident in this region.

Furthermore, Mr. Philip Chineduis a 53 year old married Igbo man with a bachelor's degree (hereafter referred to as Mr. Igantius) was interviewed in his homegarden at Ibafo, one of the satellite towns to Lagos metropolis but falls within Ogun Stateof Nigeria. The provisioning ecosystem services of his homegarden include supply of cassava (*Manihot esculenta*, F. Euphorbiaceae), medicinal herbs like lemon grass and papaw leaves. Others are vegetables like ugwu for boosting blood level in human body and pepper (*Capsicum annuum*, F. Solanaceae) which are sourced "fresh from the garden". These provided economic benefits since there were no "need to buy from the market" and social cohesion through sharing of garden products with his neighbours during the "time of economic hardship". Also, the use of ecosystem products as source of cooking energy in the cities for social gatherings are common in the region. This was confirmed during an interview in a homegarden at Omu-Aran that firewood were used for cooking to entertain guests in a wedding ceremony. Despite modernisation, fuelwood continues to be a major source of cooking energy in the region, even in the sprawling parts of small and medium-sized Yoruba cities. This is more prevalent for traditional ceremonies like burials and weddings where mass cooking is required to entertain large congregations that are usually in attendance. Fuelwood cookings are also embarked upon in some parts of the cities by citizens who have low economic profiles. For these reasons, supply of firewood is an essential provisioning ecosystem service in Yoruba cities. These practices are parts of what Cocks and Shackleton (2020, p. 249) refer to as "harvesting nature"which are crucial to biocultural relations with urban nature.

Furthermore, homegardens are sources of numerous medicinal herbs serving different purposes for wellbeing. Madam Esther cited the example of bitter leaf for the first-aid treatment of burns, mango tree (*Mangifera indica,* F. Anacardiaceae) bark and pawpaw (*Carica papaya,* F. Caricaceae) leaves for treating malaria and typhoid fevers. According to her,

> the other day, I had burns. It would have required that maybe visiting the hospital, I quickly went to the garden, got some bitter leaf, and I just squeezed, got out the bitter leaf water spread across the area where I had the burn and left the leaf on the surface. Between thirty minutes to one hour, the peppery sensation died off.

The accessibility of the herb as and when urgently needed in addition to its effectiveness in achieving desired result makes homegarden a haven for wellbeing. In confidence, she applauded this strategy that

> instead of having swelling over that place, [...] everything became normal and eventually I didn't experience peel or anything. It is even difficult for you to notice that this was what happened to me the previous day as a result of the bitter leaf I had around.

Furthermore, she identified scent leaf plant (*Ocimum gratissimum,* F. Lsabiatae) for treating stomach pain, bitter leaf plant (*Vernonia amygdalina,* F. Asteraceae) and

Fluted pumpkin (*Ugwu*) leaf (*Telfairia occidentalis,* F. Cucurbits) parboiled together, as being highly effective for treating anemia. She added that: "Before the person eats it, three, four, five times, the person is fully recovered". Pawpaw is also used to regulate blood sugar level as "it neutralizes sugar in the [*human*] system" while both fruit and leaf of soursop (*Annona muricata,* F. Annonaceae) are "very good for a diabetic patient". She addedthat "scent leaf is another antibiotic [...], it detoxifies the system, and any unwanted organism in the system, the scent leaf flushes them out".

Dr. Moses Ayoade, a Yoruba man of over 50 years of age, was a lecturer in the Department of Pharmacy in one of the federal universities in Yoruba cities. He was married with children and holds a doctorate degree in Pharmacognosy, hails from Oke-Igbo, and was interviewed in his homegarden at Ondo, in Ondo State, Nigeria. He confirmed that scent leaf (*Efinrin* in Yoruba) contains two volatile oils, "eugenol and thymol. These eugenol and thymol have a very good anti-bacteria activity and that is why [... *scent leaf*] is used as anti-diarrhea". Similar experience on the medicinal value of soursop was relayed by a Yoruba clergyman, Venerable Philemon Oginni, who was 75 years old and a native of Ikare (hereafter simply referred to as Ven. Oginni). He identified the consumption of soursop (Fig. 3.1) as a strategy for wellbeing and healthy living through preventive approach as he claimed that the plant

> has high value of medicine. It's highly medicinal. As a matter of fact, it is one of the plants in the compound [*garden*] we use the leaves every Sunday to prepare tea. I take this, that of guava, that of avocado and some others. I boil them together for the family, we put cocoa or something else and drink, and it has been helping us.

Similar view was expressed by Mr. Festus Olabanjiwho noted that soursop is highly medicinal. He explained further that:

Fig. 3.1 Soursop in Ven. Oginni's homegarden, Owo, Nigeria. Source: Photograph by author

> when you check the internet, you'll learn that it's anti-cancer, anti-aging, and it's multipurpose. In fact, we use the leaves as beverages for our morning tea. The tea has a golden colour. It's medicine for pain relief. I got the seed from my friend Mr. Awotundun.

For that reason, he declared that the family does not buy industrially produced beverages anymore, even packaged herbs like Kedi Care products. He explained that on one occasion he was bold to invite the advertisers of Kedi Care productsto his homegarden and show them the reasons why he felt he does not need such products. This kind of innovative gastro-medicinal lifestyle of IKS is another evidence that urban ecosystems are most beneficial from a perspective of multispecies ecological cities. It showcases the significance of coexistence among multispecies as established in the Garden of Eden where a complete ecological milieu provided natural strategies for the wellbeing of Adam and Eve in one and the same place (Bordeianu 2009).

In agreement with this ecological praxis (German 2004) and based on his knowledge of Western science mixed with Indigenous Knowledge System of Yoruba, Dr. Moses Ayoade confirmed that plantain has anti-diabetic activity adding that "people who have diabetics, … we encourage them to eat the unripe plantain. … If you cut it like this, juice will be oozing out from the stem, so we collect the juice and concentrate it and administer it to people who are diabetic". He explained further that Pineapple is useful for solving digestive problems like diarrhea and constipation. The peelings are boiled and sometimes with lemon and administered for the treatment of malaria and typhoid fevers. Also, Cashew "juice is used to treat trush, […] mouth trush […]. It has astringent properties. […] Some people use the leaves when boiled for the management of diabetes". He continued to emphasize that the cashew "nut contains vitamin Bs, especially Vitamin B6, Vitamin B1 which is bitamin, Vitamin B2 redoxin, […] so it serves as anti-oxidants. Anti-oxidants are anti-ageing, anti-stress, so it is very good". Also, boiled leaf of mango

> is used in the management of diabetes, then the stem bark and the leaves also, they are used in the treatment of fever […]. That is why our people do, our traditional people […], at times they can add the *Ananas comosus*, cashew, cashew stem bark and everything, they boil everything together and drink. It is very good in the treatment of malaria fever and pain. It also has analgesic activity, it helps in reducing pain and it also helps in reducing high [*body*] temperature.

He introduced the plants he called *Vernonia amygdalina* (F. Asteraceae, *Ewuro* in Yoruba name) and "*Anthocleista vogelii*, the family name is Loganiaceae. The Yoruba people call it *sapo sapo*". He continued to explain the medicinal value of the root of *Anthocleista vogelii*in the treatment of jaundice in babies and the stembark for the treatment of body pain. He pointed to another medicinal plant in the garden known as *Rauvolfia vomitoria*, family Apocynaceae and Yoruba name as *Asofeyeje*. According to Dr. Moses Ayoade, when the leaves of the plant are boiled, it is used in the management of psychiatry, making a wild psychiatric patient to sleep. He described the use of the root of the plant, which is normally long, for beating a violent psychiatric person."In the course of that […], the root will come in contact with the blood of the psychiatric patient and it will also make them to sleep. So when they sleep very well, it also calms them down", he submitted. He explained further that *Rauvolfia vomitoria* "has very good anti-hypertensive activities [… *and*] anti-diabetic activity" such that when its leaves are boiled and drank, the blood pressure and high glucose levels will reduce. These explanations suggest that Indigenous Knowledge System of ethnomedical practices through provisioning ecosystem services could be legalized as being potent for achieving human wellbeing if confirmed through scientific procedures (Eshete and Molla 2021; Agize et al. 2022).

However, Madam Esther explained that it is a challenge that "there is no particular dosage […] but when you are on it there is this caution, once in two days check your blood pressure to make sure it does not [*be*]come too low". Dosage challenge characterizes the use of herb for wellbeing (Teoh 2016) despite its efficiency in administering cure of sicknesses. Despite this, there are "so many medicinal plants, the list is endless", asshe emphasized. She argued that the homegarden has saved her from "a lot of cost, going to the hospital or the chemist, so many times", including freedom from "unnecessary panic". She identified the leaf of chloroquine plant (*Cinchona tree*, F. Rubiaceae) for the treatment of malaria fever and jaundice for infants. Accordingly, if a newborn "baby is diagnosed of slight jaundice, you boil it [*chloroquine leaf*] and give it to the baby to drink and use it to bathe the baby, [*and*] a baby that is constantly sick of malaria". Mrs. Iwapele gave similar explanation on the multifarious medico-gastronomic significance of papaw (*Carica papaya*, F. Caricaceae) for general, pre-natal, and postnatal wellbeing in IKS. According to her, the papaw fruit builds body immunity system against diseases, boiled pawpaw leaves cure malaria and typhoid fevers, unripe pawpaw fruits are used for treating jaundice in new-borne babies, and ripe pawpaw fruit aids delivery process of pregnant women to be painless. She explained that "if they[*pregnant women*] are falling into labour, they'll ask them to eat ripe pawpaw together with the seed, it'll help to deliver the baby without pains". She added that chewing pawpaw seed alone cures pile and waist pain. Dr. Moses Ayoade equally confirmed that boiled pawpaw root are administered to children suffering from convulsion while its unripe fruits are used for the treatment of pain and gonorrhea, the seed for the management of diabetes. He added that:

> people who are diabetic […] or at the pre-diabetic level, that is, their blood glucose level is close to the level of hyperglycemia between 6 and 6.5 mmol/L, so if they keep on swallowing the seeds, it helps to reduce their blood glucose level. Some people often dry the seeds and grind it too, that way, they can add it to their soup or use it to prepare tea.

For Madam Esther, another medicinal plant in IKS of ethno-obstetric post-natal care in her homegarden was Never Die plant, also called Resurrection plant, Life plant, Air plant or Miracle leaf (*Bryophyllum pinnatum*, F. Crassulaceae) (Fig. 3.2). According to her explanation on the traditional use of this plant for the wellbeing of the newly home-delivered baby is the treatment of umbilical cord. This is by adding a small quantity of table salt to a paste of the herbal leaves applied to the cord to achieve natural healing of the navel. Simply cultivated by leaf, the plant has been reported (Elufioye et al. 2022; Latif et al. 2019) to be of multiple therapeutic uses for the treatment of high blood pressure, diabetes, ulcer, inflammation, diarrhea,

Fig. 3.2 Miracle leaf in Ven. Oginni's Homegarden, Owo, Nigeria. Source: Photograph by author

malaria fever and dandruff, among many others. Similar report was given by Venerable Oginniwhen interviewed in his homegarden at Owo, Ondo State. He confirmed that Miracle leaf (*Abamoda*, in Yoruba) has many medicinal herbal values that:

> it can be used to treat cough, asthma, cold, [...], dysentery. The plant root is used for treatment of high blood pressure, [...] any cardiac problem, treatment of fever. The paste is used to treat roundworms. [...] we grind the leaf and use the extract used. [...] The juice from it can be used as ear drop; it treats diabetics. The juice for the treatment of stomach ache. It's natural treatment of certain types of cancer. It helps with great management. That is why it is called miracle leaf [...] It's miracle leaf because it cures so many diseases and so on.

Similarly, Mr. Olabanji aptly described the medicinal efficacy of miracle leaf for the treatment of high blood pressure that "If you have high blood pressure. I think our mother-in-law is taking [*consuming*] that. So, it will reduce the blood pressure. Just eat about 3 to 4 leaves". From a scientific perspective, Ingole et al. (2020, p. 162) describe Miracle leaf as "a magical herb" in view of its very high medicinal profile for the treatment of numerous diseases in many parts of the world where reliance on traditional ethno-medical practices are still very prevalent. This is because of its numerous pharmacological properties and activities including anticancer, antidiabetic, uterine-relaxant, anticonvulsant, antifungal, antimicrobial, and anti-ulcer, among so many others (Ogidigo et al. 2021). This is in view of its chemical constitutes including tannin, alkaloid, saponin glycosides, calcium, sodium, potassium, magnesium, phosphorous, iron, zinc, thiamin, and beneficial acids like benzoic acid, among so many others (Afzal et al. 2012; Ingole et al. 2020; Fernandes et al. 2021; Omojokun et al. 2021; Ogidigo et al. 2021). Saxena et al. (2003) review the scientific evidence of plant metabolites that possess anti-malarial properties in their chemical structures. They confirm the mode of action of these plants as effective treatments against malaria parasites. A recent vivid review (Sibeko and Johns 2021, p.1) of cross-cultural global traditional use of medicinal plants for ethno-obstetric purposes reveals the "antimicrobial, antioxidant, anti-inflamatory, immunomodulatory, antidepressant, analgesic, galactagogic, and safety properties" of "1948 species, 105 plant taxa".

Mint leaf (*Mentha spicata*, F. Lamiaceae), star fruit (*Averrhoa carambola*, F. Oxalidaceae) and chestnut (*Castanea sativa*, F. Fagaceae) are other gastro-medicinal plants in the homegarden of Ven. Oginni. He explained that chestnut can be processed by frying or sundrying before removing the peels and then ground. The ground form is used for culinary purpose as thickener for traditional soups. Aside from this, the level of IKS for wellbeing through homegardens in the region can only be described as being amazing. It appears as if most of the owners of homegardens are specially trained in ethnobotanical strategies going in view of the details of their expository assertions with sureness. This notion is clearly manifested in the words of Ven. Oginni when accounting for the provisioning ecosystem services of star fruit plant (Fig. 3.3). He expounded that:

> as it is, why is it called star? You can see that the shape is irregular. When you cut it [...], it will just look like star. It has so many uses. It can be used as salad. You can juice it or blend it. Vitamin C content is about 76%. It washes off cold, flu, and any other type of infection. It prevents constipation. It helps in loss [*of*] weight. It helps to prevent the absorption of cholesterol in the body. It helps to neutralize harmful free radicals that can cause inflammation in the body. As a matter of fact, people who have diabetes also use it and it reduces the effect of diabetes in their body. So, we were told that the leaves and the fruits have been used to stop vomiting.

Fig. 3.3 Star fruit plant in Ven. Oginni's Homegarden, Owo, Nigeria. Source: Photograph by author

Furthermore, the leaf of Guava plant (*Psidium guajava*, F. Myrtaceae) was identified as being medicinal for treating dysentery, sugarcane (*Saccharum officinarum*, F. Poaceae) for recovering fromdehydration, avocado pear for high blood pressure, and moringa (*Moringa oleifera*, F. Moringaceae) (seed, leaf, stem) for different ailments including high blood pressure. Dr. Moses Ayoadeequally submitted that guava has anti-oxidant properties because it is rich in vitamin C and other vitamins. The multiple medicinal herbal use of Levant cotton (*Gossypium herbaceum*, F. Malvaceae, *Ewe owu* in Yoruba name) was explained to be due to its components including "gossypium and emigossypium which have male anti-fertility activity" and therefore used as male contraceptive. "If the seed is boiled and drunk and ingested by men, it will sterilize their sperm", he emphasized. According to Dr. Moses Ayoade, the major ecosystem disservice why this plant has sideeffectis that "it also has nephrotoxic activities, that is, it has ability to destroy the nephron, the kidney cells". Again, this buttresses one of the challenges of ethnobotanical medicine and strategies for human wellbeing in Yoruba ecological urbanism. Conversely, Prof. Akinradeyo, interviewed in his homegarden at Ogbomoso, submitted that "potato is highly medicinal for men" for increasing sperm count. This claim has also been confirmed in the literature through narratives of life experiences in other climes (Swan and Colino 2022).

In an interview with Mr. Folarin A. S.. in his homegarden in Akure, he emphasized that moringa seed can be eaten, its leaves could be boiled, and the tree bark could be soaked in water as herbal concoction for medication. Moringa also has sedative properties. This was confirmed by Ven. Oginni who explained that when he was working as a priest at Ijebu, another Yoruba city in Ogun Stateof Nigeria, that he had a friend who visited him at a time and offered him the moringa fruit which helped him as sedative. In an onsite interview with Dr. Moses Ayoade in his homegarden, he confirmed these claims from scientific perspective. He explained that moringa has "anti-diabetic, anti-ulcer, anti-rheumatic, anti-bacteria, anti-hypertensive and so many more health benefits and the reason is that it is very rich in vitamins – C, B1, B3, B4, B12. So, it has so many anti-oxidant activities and also works in preventing cancer".

Furthermore, while underscoring the sedative property of moringa seed (Fig. 3.4), Ven. Oginni added that "if you cannot sleep, you will sleep properly. People eat it raw as it is; then you can add to another ingredient". It is pertinent to point out that ethnobotanical ecosystem services in the region are not only curative but also preventive. The practice of preventive medicine is a cultural lifestyle that values ecological relationship of humans with the world of nonhuman nature. A vivid example of this IKS was described by Ven. Oginni how the entire family members live on King of bitters plant (*Andrographis paniculata*, F. Acanthaceae) that looks like scent leaf. According to him, this is"what we do in the family [...]. So, once a week, [...] each of us takes a glass [*cup*] and it takes care of us for the whole week. By the grace of God, we and fever, we are not friends in this family" because "it is antitoxin. And it can be used for the treatment of infection and some diseases. It has beneficial effect on the liver". These claims on IKS for preventive consumption of the two medicinal plants have been confirmed in applied science. For instance,

Fig. 3.4 Moringa plant in Ven. Oginni's Homegarden, Owo, Nigeria. Source: Photograph by author

Sharma and Sanadhya (2017) report on King of bitters that it has "anti-inflamatory, antibacterial, antiangiogenic, antioxidant, hepatoprotective, antimicrobial, anticancer, antivenom, anti HIV, antimalarial, antipyretic, antifertility, antidiarrhoeal, antidiabetic, antihiperlipidemic, and antitoxicity activities" (p. 117). Similar pharmacognostic reports have been given by other scientists including Casamonti et al. (2019), Richard et al. (2017), Mehta et al. (2021); Parveen et al. (2019), and Hossain et al. (2021), among many others, largely claiming that extract of the plant has no allergic reactions that are anaphylactic.

Water leaf (*Talinum fruticosum*, F. Talinaceae) is another homegarden plant in IKS of Yoruba that is considered as a vegetable. It is laden with vitamins, among other large number of anti-malaria plants discovered to be potentin the ethnobotanical lifestyle and ecological urbanism of Yoruba people (Dike et al. 2012). Furthermore, provisioning ecosystem services of the homegardens in the region transcend medicinal herbal plants. They also include numerous gastronomic vegetables, fruits, roots and tuber crops, a form of urban crop gardening and agriculture for food production. This has been argued to be 'more beneficial' and therefore embraced than ornamental homegardening. In the homegarden of Mr. Olagunju, there was only one ornamental tree which was planted for the purpose of providing shade and thermal cooling effect for the rabbit space. The idea is hinged on an IKS of Yorubathat food is medicine, just like the wild and cultivated medicinal herbal plants.

In this homegardening genre, fruits like pineapple (*Ananas comosus*, F. Bromeliaceae), passion fruit (*Passiflora edulis*, F. Passifloraceae), cashew (*Anacardium occidentale*, F. Anacardiaceae), sweet orange (*Citrus sinensis*, F. Rutaceae), and plantain (*Musa paradisiaca*, F. Musaceae) are common examples. This is with the understanding in IKSthat all fruits are sources of vitamins for building the immune system of the body, including plantain that is a good source of

Iron in particular, according to Madam Esther. She argued that "if you want to be strong", fruits are indispensable. Likewise, to emphasize the significance of homegardens as sources of fruits for wellbeing through IKS, Mr. Olabanji stressed that fruits build the immune system of human body against diseases. According to him, regular intake of fruits from the garden enhances agility and resistance to sicknesses. He drew allegory from the Coronavirus disease to which some people in contact with infected persons were not infected while others were infected adding that soursop in particular also has anti-aging constituents. In another scenario, Ven. Oginni confirmed that Passion fruit "has high vitamin C content and [...] there was a time [*his daughter*] was having toothache and I [*he*] gave her, when she drank [*it*], that was the end of the ache permanently". He added that Ivy gourd (*Coccinia grandis*, F. Cucurbitaceae) (Fig. 3.5), in his homegarden was highly medicinal as "it improves the [*human body*] metabolism; reduces blood sugar level; and it is good for the heart [*functioning*] and the nervous system; it can be eaten raw, that is, the fruit". It improves the digestive system and medicinal for skincare, based on the experience of Mr. Olagunju. However, according to him, the challenge of ecosystem disservice of Ivy gourd is the susceptibility of its ripe fruit to reptile invasion, especially lizard that is most prevalent in the region.

Ven. Oginni described Jackfruit (*Artocarpus heterophyllus*, F. Moraceae) in his homegarden at Owo as being highly beneficial, serving both medicinal and gastronomic ecosystem services. According to him, "first and foremost, it is food; proper food. When it is fully ripe, you cut it, there are seeds inside. You can dry it. You can dry it and eat it raw as it is. Then the flesh inside is also good; it takes care of so many things [*ailments*] in the body and it is very good". Also, the stem juice of the Musaceae family are generally used for first aid treatments of human blood bleeding due to cuts as clotting agent. There are also plants grown in homegardens that serve both medicinal and gastronomic purposes through culinary process. Tea leaf (lemon grass) (*Cymbopogon citratus*, F. Poaceae) was mentioned to be widely

Fig. 3.5 Ivy gourd plant in Ven. Oginni's homegarden, Owo, Nigeria. Source: Photograph by author

known in this category as it is used to prepare soup and "for the treatment of [...] very many ailments that the list is inexhaustible". An important gastronomic plant that also doubles as medicinal herb is round pepper (*Capsium fructescens*, F. Solanaceae, Yoruba name *Ata rodo*) because it contains "a chemical, Capsaicin". According to Dr. Moses Ayoade,

> this Capsaicin has a very good anti-arthritic activity. It also has neuroglial activity, that is, people [*who*]have pain in their neurons and inflammation pain, arthritic pain, rheumatic pain. [...] It is what they normally use. They combine it with cream and apply it to the body of people suffering from arthritis.

Ven. Oginni revealed how he adds personally-prepared cocoa powder to his tea drink because the cocoa plants (Fig. 3.6) are available in his homegarden. He remarked that, "cocoa is very good for the body; when you drink it, it gives you freshness and it makes you to be very vibrant. As I was telling you earlier that when [*I*] prepare those tea, I add it to cocoa powder. I prepare it myself". The availability of fruit plants in homegardens is a deterrent from the consumption of packaged fruit juices. For instance, the assorted fruit plants in Ven. Oginni's homegarden assured a continuous daily living with the consumption of fresh fruits like Sweet orange (*Citrus sinensis*, F. Rutaceae), (Fig. 3.7). The consumption of raw unprocessed food items, including beverage plants, has been appraised to have better health outcomes compared with industrially processed food items (Salomé et al. 2021).

Dr. Moses Ayoade provided further insights into the link between science and IKS on ethnobotanical worldview of the Yoruba people. He introduced a medicinal-culinary herb in his homegarden that he called it Icacina plant (*Icacina tricantha*, F. Icacinaceae). He proceeded as follows:

> This plant is called *Icacina tricantha*. Yoruba people call it *Ewe gbegbe*. Normally, they use it to wrap beans cake. [...] but it is used traditionally in the management of inflammation, swelling in any part of the body, and arthritis. So it has a good anti-arthritic and anti-inflammatory activity and it also has a good analgesic activity for treating body pains, yes for rheumatism.

Fig. 3.6 Cocoa tree in Ven. Oginni's homegarden, Owo, Nigeria. Source: Photograph by author

Fig. 3.7 Sweet orangetree in Ven. Oginni's homegarden, Owo, Nigeria. Source: Photograph by author

Similarly, while lime (*Citrus aurantifolia*) and lemon (*Citrus limon*) trees (both in the F. Rutaceae) produce tender leaves for culinary ecosystem service, the fruits are used for numerous medicinal purposes. In addition to the herbal use as anti-malaria, there is a widely reported understanding in the region that slicing and boiling lime or lemon in water and drinking such in the morning makes the stomach wall to be alkaline. According to Madam Esther, this wellbeing strategy is claimed as means of purifying human body from the environmental pollution of smokes from automobiles and electric power generators highly prevalent in Yoruba cities. Similar lifestyle of consumption oflime and lemon for wellbeing was detailed by Mr. OlabanjiHe declared his indigenous knowledge that juices of lime and lemon are alkaline and enhance the building of"body immunity [*system*] and resistance to diseases". In agreement with these claims, Dr. Moses Ayoade aptly gave scientific insight into the anti-malaria activity of lime that "some people also use it in the treatment of cough" boiling "it with other plants like *Dongoyaro* and *Ananas comosus*" to drink.

The combination of plant cultivation and animal husbandry in many Yoruba homegardens is an eloquent testimony that a complete ecosystem of flora and fauna, including food web is practised. Aside from other manifestations of this lifestyle in traditional religious ritualism, this practice attests to the value-system of Yoruba people on multispecies ecological urbanism. The ecological passion for this form of IKS ofurbanism is matched with ease of plant cultivation due to the highly favourable environmental conditions. Arguing this point, Madam Esther explained that "I did not do any special planting. What I did was when I consumed any species I like, I deposited the seed at the place where I wanted it to grow, then when I came back after a week or two, I saw that they were already growing". The reason for this was explained that "the land here is moderately good. No special thing, we just planted and it started growing". Indigenous knowledge of agricultural practices are central to the achievement of the United Nations Sustainable Development Goals on food security, ecosystem health, and ecological sustainability (Lal et al. 2021).

Accordingly, the homegardens are also widely believed by the Yoruba people to be sources of healthy living. This importance is attached to gastronomic materials that are not only fresh but accessible in timely manners. Mrs. Faremi bared her mind on this gastronomic ecosystem servicewhen she claimed that:

> growing my own fruits and vegetables helps me to eat healthier. All I want from the [*home*] garden are always healthier. I eat fresh vegetables, fruits like pawpaw, cashew, [*and*] mango. I have enough vegetables like pumpkin, bitter leaves, scent leaves, aloe vera, tomatoes, sugar cane, lemon grass, guava, and sometimes when I have malaria, I gather the leaves together

Healthy living and human wellbeing have also been associated with dietary lifestyle in science (Carrus et al. 2018; Liu and Grunert 2020). An important aspect of this is related to the extent to which dietary contents are 'natural' or 'artificial' (Ares et al. 2014). Mr. Afolabi, who holds HND in Electrical Electronics Engineering, is a middle-aged man ancestral to Akoko area but maintains a garden in his home at Akure, both in Ondo Stateof Nigeria. He related the significance of his homegarden to the larger debates on the relationship between dietary quality and human wellbeing. He emphasized that he sources fresh food items from his homegarden:

> [...] fresh from the source, not the one they shipped [...]; this one is natural. That is why they always advise individuals to just get a small garden [...] within your environment [...], and you will not even apply all these kind of fertilizers, no, this one is natural.

The Biblical Garden of Eden vividly illustrates an environment for both humans and nonhumans and an ideal ecological milieu for human wellbeing (Clewell and Aronson 2006). The garden was a natural abode for human Adam and Eve, the non-human animal and plant kingdoms, along with illustrious waterscapes of its four rivers (Stone 2019; Kang 2020). According to Mr. Afolabi,

> All these tomato pastes of a thing is packaging, they will add preservatives but these ones are natural [...]. Just imagine the olden days; look at our forefathers, everything from farm is natural, and they always lived long, because that thing will just go into their system naturally, fresh pepper, fresh tomato, fresh vegetable, fresh oil from the palm. You see them, they don't easily fall sick, before you know it, if they started feeling headache, they'll just go into the garden, get some leaves, and grind them together and drink it; they'll say herbal medicine [...]; all these Panadol whatever, where do you think they are sourced from?

Therefore, it was not surprising that when Mr. Olabanji narrated that his dream of establishing homegarden was to replicate the Garden of Eden. In view of the ecological serenity of his homegarden, he recalled his interaction with his friend: "In fact, my friend calls it Garden of Eden. [...] somebody came here for the first time [...], he said, 'this is the garden of Eden'. I said 'have you been to Eden?' He said he has a picture. So, [...] even the animals are there". The mental creations of the Garden of Eden by Mr. Olabanji and others that led to the creation of his homegarden are laden with psychological truths. Their perceptions are explicable as brilliant representations of physical realities constructed in human minds and imaginations.

These imaginations exemplify human attachments to places that are built on the creation story eco-theologically. A vivid text on the Garden of Eden in relation to horticultural tastes is presented by Solomon (2020, p. 15) who argues for "comparing real and literary descriptions of nature" (p. 773) and connects "the material and immaterial in a way that animates the landscape" (p. 787). In another sense, Mr. Abiola was very proud of the provisioning ecosystem services of his homegarden on how he perceived it to have contributed to his wellbeing, and slowing down his ageing process. He was of the opinion that his homegarden enhances "[…] consumption of fresh and natural nutrients, […] now I'm 45 [*years old*], I'm still very young", he remarked. This kind of shared personal experience on self-perceived wellbeing and longevity of life based on home environmental quality and dietary system has been established elsewhere (Baker and Steemers 2019; Longo 2018).

The actors on medicinal urban gardening in Yoruba cities are not limited to any educational status nor socio-economic group, generally. Mr. Hamzat Olukoye, a married Yoruba, who hails from Ikare Akoko in Ondo State, was the supervisor of Odugbemi Medicinal Garden Oke-Igbo, also in Ondo State (hereafter simply referred to as Mr. Olukoye) During an interview, he revealed that "this [*the*] garden started around 2008, twelve years ago, by Professor Tolu Odugbemi with one of our fathers, Professor Daramola […] with the wife of my boss, Mrs. Titilayo Odugbemi, with me and one of my friends Mr. Odunayo". The incredible sticking to IKS by highly educated elites in Yoruba cities is remarkable. To show that this is not a taken-for-granted easy affairs but requiring great efforts in plant sourcing, Mr. Olukoye explained that "some of these [*the*] plants were sourced from northern area [*of Nigeria*] from Calabar too, towards Maiduguri area, Kaduna, Ilorin, Ikare-Akoko where I hail from, Oke-Igbo here, that's Igbo-Olodumare [*forest of God*], and Cameroon". This explicitly means that the plants are mostly ex-situ from across the whole of Nigeria and other country likeCameroon. *Igbo-Olodumare* is a sacred grove cultural landscape of extremely high biodiversities in Oke-Igbo, from which the name 'Oke-Igbo' was derived.

Furthermore, mystic ecology is an important provisional ecosystem of the medicinal gardens in Yoruba urbanism. An example of this mysticism was a plant shown by Mr. Olukoye which he named "*Eyin Loju Awodi*" or "*Eyin Olobe*" (*Phyllantus amarus*) and described as anti-bullet. According to him, "herbalists say if you swallow three seeds, if you are shoot with gun […], the bullet won't enter your body, and cutlass too". Another plant of similar mystic use in the garden was "*Kansinkansin*" which "if snakes pass it […] the snake can die" and "king of bitters (Fig. 3.8) […] which suppresses high blood pressure […]. It is eaten directly too. This one is *Ogbo* herb. It's an anti-malaria drug. It's the one our [*fore*]-fathers used to consume" for their health wellbeing, he explained. The claims on anti-bullet and anti-cut charms in Yoruba mysticism and superstitions have been researched in scientific praxis (Olawa et al. 2020) with the conclusion that they are efficacious in fetish practices.

The metaphysical complexities of these practices and the influence of non-traditional religions, especially Christianity, have made them limited to traditional actors for self-protection. This is especially in frontline, mainstream and opposition

Fig. 3.8 King of bitters plant in Ven. Oginni's homegarden, Owo, Nigeria. Source: Photograph by author

politics, thuggery, robbery, hooliganism, and other negative uses by anti-social actors in the process of securing state-criminalized alternative survival strategy in an over-stretched polity (Ajibade 2006). Similar claim of mystic ecology was made by Mr. Olabanji who identified resurrection plant as being anti-evil. Claiming that he sourced the plant from a sister and emphasized that resurrection plant "is also called back-to-sender. For those who knows it, they said if you plant it inside your house, no evil can come near your house". Kelly (2018) described this form of mystic relationship with nature as a form of spirituality that has capacity to explain the "connection between spirituality and ecology" (p. 181).

IKS of food packaging is rampant among the Yoruba people from pre-colonial time to-date, even among the ranks of literate elites. Dr. Moses Ayoadeexplained that Sandpaper plant (*Ficus exasperata*, F. Moraceae, *Ewe eepin* in Yoruba) is a leaf for culinary maintenance of cooking pots in IKS instead of using industrial scouring powder. He explained that the same plant has ethnomedical use as he explained:

> This leaf is very good in treatment of gonorrhea. People who have gonorrhea when they boil it and they drink it, it cures gonorrhea [...]; I know some people also use it for the management of diabetes in some parts of the country.

Other forms of ecosystem services are culinary practices thatexplore special leaves of wide surface areas, required strength and stiffness for wrapping food items which taste better than food wrapped in synthetic polythene. The most popular of these leafy wrappers are called *Ewe-eran* (*Sarcophrynium brachystachys*) and *Ewe-gbodogi* (*Megaphrynium macrostachyum*). *Ewe-gbodogi* was found in Fagbemi Medicinal Plants Garden at Oke-Igbo. The two types of culinary leaves are not only used for domestic purposes but also in high-caliber public social events like wedding, birthday, and burial ceremonies. *Osunsun* (*Carpolobia lutea*) was another important medicinal plant in that Garden. It serves multiple purposes as aphrodisiac herb for "curing male sterility, increasing libido, enhancement of aphrodisiac prowess, enhancement of virility and male fertility, and augmentation of male sexual functions" (Nwidu et al. 2015, p. 133). In a symbolic relationship with the aphrodisiac use of this herb, Mr. Olukoye reveals the importance of its stem as beating cane in the African mode of child training forpunishing offending children in schools. He recalled his personal experience that "anytime they want to beat us, they will say go and source *Osunsun* for me". Furthermore, despite civilisation and modernisation, the connection of Yoruba people to some traditional modes of hygiene is still intact. A clear example is the herbal plant called *Pako-Ijebu* (*Massularia acuminata*) that is used as chewing stick. *Pako-Ijebu* has been discovered to contain Sodium ion, Calcium ion, Zinc ion, and Potassium ion (Yakubu et al. 2008). It has also been proved to be an excellent strategy for dental hygiene because of its antimicrobial activities (Taiwo et al. 1999).

Yoruba people also derive ecosystem services from green infrastructure for recreational purposes. The development and test of intelligence through traditional games has been adjudged to be a unique feature of Yoruba human ecology. This provisional ecosystem service is enhanced with the "*ayo*"seed plant, Gray Nicker (*Caesalpinia bonduc,* F. Caesalpiniaceae) whose leaf is also "used as medicinal plant for measles" according to Mr. Olukoye. The popular outdoor game called "*Ayo Olopon*" is a socio-cultural high cognition recreation game of the Yoruba people. The game board has 12 holes in two rows with four *Caesalpinia bonduc* seeds in each hole to make a total of 48 seeds for the two players. The game is not only an ecological symbol of the Yoruba people, it has been associated with the Yoruba value-system, native intelligence, cosmology, traditional religions, and philosophy (Òkéwándé 2017). Currently, it has been internationally popularized with different names through globalization processes. Also, all parts of *Caesalpinia bonduc* plant, including seed, root, and leaves have been found highly medicinal and grown in homegardens in many other countries of the world. These parts are used as medication for stomach disorders, snakebites, cancer, "piles, intestinal worms, elephantiasis, splenomegaly, hepatomegaly, amenorrhea, dysmenorrhea, pharyngodynia, and fevers" (Akbar 2020, p. 483).

The medicinal value of *Osun-buke*, Achiote (*Bixa Orellana* Linn, family Bixaceae) as "phytocosmetics in traditional medicine orally passed down through generations in South West Nigeria" (Fred-Jaiyesimi et al. 2015, p. 313) was emphasized by Mr. Olukoye at Odugbemi Gardens. Leaf and seed parts of this garden plant have also been scientifically verified as medication for dermatological disorders like rashes, pimples, eczema, dis-colouration, among many others (Goyal et al. 2022; Paniagua-Zambrana et al. 2020; Siriyong et al. 2020). Aloe vera (*Aloe barbadensis,* family Xanthorrhoeaceae) was another very important medicinal plant in the Garden. According to Mr. Olukoye, Aloe vera gel is another phytocosmetic product for the treatment of "germs on the body [...*causing*] scratching". It has also been reported in science that *Aloe vera* has many medicinal properties including "immunomodulatory, wound and burn healing, hypoglycemic, anticancer, gastro-protective, antifungal, and anti-inflammatory properties" (Maan et al. 2018, p. 1). Coconut (*Cocos nucifera*, F. Arecaceae) is another important plant in Yoruba cities. Mrs. Iwapele gave insight on its ecosystem services that the fruit is usedfor skincare in IKS. She emphasized that "the hard bark can be used to cure rashes when boiled with water and used to bathe". Also, locally-extracted coconut oil cream is suitable for the newly-born "instead of buying artificial hair cream and body cream that may cause rashes on the body". According to her, coconut fruit water mixed with honey relieves cough.

Furthermore, Calabash tree (*Crescentia cujete*) was the source of traditional utensils in Yoruba culturebefore the advent of modern utensils. This was achieved by removing the pulp of itsdried and sliced fruits. The leaves of the plant also contain phytochemicals which make it suitable for numerous medicinal purposes not only in Yoruba cities but also across the Global South. The ecosystem services of Calabash tree are also manifested in the use of the fruit pulp as sheep feeds (Islam et al. 2019). Mr. Olukoye identified the provisioning ecosystem services of bitter leaf and bark of *Dongoyaro* (*Azadirachta indica*) as medicinal, being antimalaria plants. Malaria is a highly prevalent arboviral disease in the tropical climate of the Yoruba cities. However, the availability of these medicinal plants in abundance has proved to be reliable succor in this regard for the wellbeing of Yoruba people.

3.3 Supporting Ecosystem Services of Yoruba Home Greens

Homegardens and green spaces play crucial supporting roles for wellbeing in Yoruba cities. This category of ecosystem services has been confirmed to be beneficial to the human and nonhuman ecosystems of Yoruba cities. Mrs. Faremi explained that her homegarden:

> helps to give a healthy dose of vitamin D. In the morning, when I go to the garden to work, I gain vitamin D from morning sun. Whenever I go to the garden to weed or do any kind of gardening activities, I will go in the morning and work till the morning sun rises which gives vitamin D and as a matter of fact, vitamin D helps older adults like me.

This signifies the ecological support of the urban ecosystem by the abiotic elements like the sunlight as a primary source of planetary energy system. In agreement with these claims, the synthetization of the naturally occurring vitamin D3 called cholecalciferol in the skin of older human population through the ultraviolet radiation has been specifically confirmed in science (O'sullivan et al. 2017).

Homegardens are significant for the support of terrestrial and aquatic food webs of the urban ecological system, although some aspects of this have been branded as ecosystem disservice in Yoruba cities. Mrs. Faremi buttressed this while describing the installation of scarecrow in her homegarden. According to her,

> scarecrow […] means arranging different sticks to form an image that looks like a man standing in the garden with some bells on its body and when the wind blows, it makes noise and scares the animals away.

She explained further that this strategy becomes a necessary solution to the ecosystem disservice that arises from supporting ecosystem services in the foodwebas:

> rodents attack roots, stem, leaves, fruits, and seeds of the plants. […] Birds peck out seeds when they have been planted, they eat young plants and fruits. Crabs destroy vegetables in wetland of the garden during dry season. Insects also destroy crops by eating their fruits,

> leaves, stems, and roots. Grasshopper and locust eat the leaves of plants. Termites eat the stem and roots of plants.

In a similar way, Mr. Abiola reiterated that "immediately you plant the maize like this, all these rabbits will come and eat it". Similar ecosystem disservices were experienced elsewhere in the homegarden of Mr. Philip Chinedu where "domestic animals like goats and other animals, even some that prey in the night like [*wild*] rabbit" eat up homegarden crops. He decried that plants like maize were uprooted by these rodents and foliage eaten up by domestic goats until perimeter fence was erected to effect territorial control of the homegarden. Scarecrow has been affirmed to be a potent cultural strategy in IKS to minimize the ecosystem disservice in food webs (Wilk 2014).

Similar ecosystem disservice was decried by Madam Estherin her homegarden as she explained that she has "so many unwanted animals creeping in to destroy" her plants with an example of small wild snails that feed on her *ugwu* plants. This made her to resort to IKS of sprinkling ashes on the plants in order to discourage the animals from foraging on them. Despite such unwanted ecosystem disservice necessitated by the natural food web, she appraised nature by emphasizing that some of her plants were not propagated by her but by natural seed dispersion mechanisms. The naturally dispersed seed would normally grow only during the rainy season. In view of absence of water for the growth of plants in the dry seasons, manual irrigation during the dry season is a common homegardening activity in the Yoruba cities. Dr. Moses Ayoade narrated his own experience that he waters the garden early in the morning. He also described the generosity of 'nature' that allows freely fallen seeds to germinate and spring up speedily immediately at the onset of rainy season. To describe this generosity in stimulating an ecological urbanism of multispecies, he cited how:

> God blesses the soil, [*as*] it brings forth fruits to provide food for both [*hu*]mans and animals like birds but one thing we do minimize incidence like that, is that we position scarecrows at strategic places in the farm [*garden*] to scare away animals

However, Mrs. Folagbade in her own case described the ecosystem disservice in her homegarden as unbearable, necessitating the occasional application of Gammalin 20 pesticide for termite fumigation. In an extensive review of the literature on the multiple implications of pesticide use in Nigeria, Ojo (2016) identified "disruption of ecological balance and collapse of biodiversity" (p. 981) as parts of the negative outcomes in the ecological urbanism in Yoruba cities. He emphasizes that chemical reliance is not a sustainable practise for the ecosystem. Mr. Philip Chinedu equally denounced the application of "pesticides or anything that can affect the crops" but rather engaged in manual garden management procedures in the pursuit of organic crop production.

Furthermore, the ecological urbanism of the Yoruba people presents a balance between human and nonhuman nature, flora and fauna components. Accordingly, as multispecies settings, the homegardens are not green spaces only, they harbour series of animal husbandries. Mrs. Folagbadewas interviewed in her homegarden in Akure on the commitment to animal husbandry. She confirmed that her family rears dogs and maintains poultry section in the homegarden. Also, Mr. Abiola describes this mixed-husbandry practice of gardening in his homegarden as follows: "there's a goat [*garden*], there's piggery, there's broilers [...], we have a dog, we have other things; had it been you come [sic] during December 23, 24; now we have disposed of most of our birds". He emphasized that he relies on organic manure system as follow: "I don't use fertilizers, I can only use all these wastes from the birds, pigs as you can see". This form of integrated gardening provides supporting ecosystem servicesof nutrient cycling and biological nitrogen fixation being symbiotic and ecologically self-sustainable in terrestrial ecosystems (Zheng et al. 2019). Ven. Oginni explained the significance of supporting ecosystem services of homegardens through complementarities of flora and fauna ecosystems. To achieve sustainable thermal micro-climate for the aquatic ecosystem of the fish pond in his homegarden, he planted plantain adjacent to the pond. This combination enhances environmental cooling effect by shielding the pond from high solar insolation that is prevalent in the tropical climate.

Conversely, Ven. Oginni claimed that he "deliberately planted the plantains [*near the fish pond*] so that the water from my fish pond will take care" of the plantain constantly and whenever the pond is drained. The homegarden of this priest at Owo can be described appropriately as urban ecological garden with many supporting ecosystem services. The garden has coat bottons plant (*Tridax procumbens*, F. Asteraceae) that the family members consume as vegetables and he adds to the feed for the domestic rabbits (*Oryctolagus cuniculus*, F. Leporidae) he rears in the same homegarden. Equally, Mr. Olabanji advanced another dual purpose of leaves of soursop, avocado, pawpaw, and guava as the "morning tea" that the family members "boil and drink" as herbal beverage for medicinal reasons and feed the rabbits in the garden with some of the leaves. The ecological co-existence of flora and fauna in homegarden was also amazing as Turkey (*Meleagris gallopavo domesticus*, F. Phasianidae) feeds on mango (*Mangifera indica*, F. Anacardiaceae) and tangerine (*Citrus reticulata*, F. Rutaceae) leaves. As an erudite scholar, even at the high status, Professor Akinradeyo also practises multispecies homegardening in Ogbomoso, Oyo State of Nigeria as he claimed:

> That's why I support it with birds. The geese are there. I have the female and the male goose. Very soon, I will bring the goats and other animals there. Actually, that place will be my animal garden and this place will remain permanently vegetable garden

These practices of mixed-husbandry homegardening have been found to support human livelihood, urban ecology, and urban biodiversities (Nolan-Spohn 2005; Pushpakumara et al. 2020). Experiential narratives on urban ecologies of the interconnection of supporting ecosystem services among multispecies in homegardens are eloquent testimonies that complete urban ecosystems are most functional for humans and nonhumans (Šiftová 2021; Ghimire and Joshi 2020). These human-nonhuman connections for small-scale food production in urban home environments are crucial to the achievement of the Sustainable Development Goals and Agenda 2030 of the United Nations (Nicholls et al. 2020). The provisioning ecosystem services through the urban horticulture homegardening translate into supporting ecosystem services that result into urban sustainable food system, human health benefits and wellbeing (Jahrl et al. 2021; Nogeire-McRae et al. 2018; Cruz-Piedrahita et al. 2020). The quality of homegarden soils is central to the realization of these multiple benefits of urban ecosystem services. Among the case studies conducted in the study meta-narrated in this book, Mr. Olukoyepointed out that his homegarden has different types of topsoils in different parts of the garden. In this type of scenario, the different categories of topsoils would support the growth and yield of different food crops (Raj et al. 2019; Wood et al. 2018).

3.4 Regulating Ecosystem Services of Yoruba Home Greens

Urban homegardens of Yoruba people are regulatory ecological landscapes. The regulating ecosystem services of these homegarden landscapes are enhanced through multiple and complex combinations of waterscapes, drylands and wetlands. They are rich urban ecological zones that are adaptable to climate variations, sustaining the quality of soil nutrients, mineralization of soils through adaptive use, regulating micro-climates, controlling soil erosion and excessive runoff waters, and adapting the soil to cultivation of multiple consumable crops. The comments made by Mrs. Faremi corroborate this claim. She explained that:

> but this plot was full of water, it is a swampy area, because there is a stream behind my compound, which makes my garden to be wet and swampy every raining season. This made me to make little drainage in the compound that leads to the back of the compound.[...]. This garden part of my compound was sloping, so erosion use [sic] to erode the garden easily. To prevent that, I made drainage, then I noticed that with the drainage, the erosion will still erode the minerals in the soils, so I decided to plant some plants that will reduce the erosion when it is raining. I planted the two species of scent leaves, aloe vera, lemon grass, and we left grasses in some part of the garden to control erosion.

The role of plants in preventing excessive surface water runoff, sheet and rill erosion, and eventual nutrient loss cannot be overemphasized. Natural land covers and accompanying plant roots play essential roles in the conservation of the earth's natural capital (the soil), and soil ecological processes and stability are foundational to all other ecosystem services (Ruiz-Colmenero et al. 2013; Woznicki et al. 2020). All these ultimately lead to high efficiency of homegarden soils to support the sustainable growth of quality green spaces that regulate micro-climate of home environments towards human productivity and wellbeing. In consonance with these propositions, Mrs. Faremi pointed out that:

> This garden also helps me during the heat period, because inside the house can be hot and there might not be light [*power supply*], so I will just come to the garden with my chair and sit under a tree. I receive fresh air and shade from there.

This experiential regulating ecosystem service is not an isolated case but a common aspect of IKS of urban green infrastructure in the region. Mr. Olabanji equally explained that the trees "shade the house from the heat of the sun" and direct solar insolation is avoided on the envelope of the building with resultant natural cooling effect on the interiors of the building and "freshness of the air". To emphasize the outdoor-indoor significance of the trees in his homegarden, he explained further as follow: "In fact, you see the bench outside there, some people will come and sit outside instead of sitting inside, under the tree. We relax there. I put my mattress there on the bench and I sleep off". He added that the homegarden also prevents soil erosion and adds ornamental beauty, serenity, and conducive thermal effect to the environment. Indeed, the green spaces of the home added ornamental aesthetics, as a cultural ecosystem service, that made the home environment unique.

The natural canopies of the trees protect from inclement weather, and ensure conducive indoor thermal quality of the building by providing cooling effects. Mrs. Iwapele also affirmed that her homegarden was meant to provide "shade in the house" for relaxation under trees as a form of passive recreation. These nature-based solutions to urban environmental challenges have been praised to be sustainable, effective, and cheap, with attendant enhancement of human wellbeing (Santiago Fink 2016). All species of humans and nonhumans benefit from these regulating ecosystem services. To Mr. Afolabi, his homegarden is a cool place for both human and nonhuman fauna. According to him:

> Even the piggery [sic], they like a very cool environment. If you come here in the late evening, you'll like the breeze, at times you'll not even see me at the sitting room. I will just go to the garden and be strolling, taking the fresh air. Sometimes, I may like to make a call, I will go into the garden there and be making call, receiving call, nobody to distract unless the breeze of the whole crops [*which*] is natural.

This form of passive recreation involving strolling, sitting, and contact with nature are core ecological regulating ecosystem services that have been discovered to enhance human wellbeing (Talal and Santelmann 2020; Herman et al. 2021; Roe and McCay 2021).

Also, to provide support for the maintenance of the home environment and continued access to the ecosystem services of the home greens, Mrs. Faremi disclosed that "I made drainage [...]. I made channels for easy flow of water down to the stream behind the compound". In some of the homegardens investigated in this study, the wet hydrological quality of the soils in some portions of the gardens is explored for aquatic ecosystems of natural fish ponds. Such hydrological profile is due to plot locations partly or wholly in or in the proximity of water catchment areas or excessively high levels of ground water-tables. This method makes the gardens to drain properly and supports their ecological qualities. The combination of green-blue spaces in these homegardens doubles for the purpose of micro-climatic modifications of the homes for indoor thermal comfort. However, they also have negative outcomes and challenge of ecosystem disservice of mosquito breeding in un-fumigated gardens with attendant high risk of malaria infections to the home dwellers (Matthys et al. 2006; Swanson et al. 2009).

The significance of homegardens in providing regulating ecosystem services of wind speed breaking cannot be overemphasized. While explaining the importance of his homegarden, Mr. Abiola pointed out that:

> another use of this garden is that some tree plants like mango, cashew, and other trees help to break wind speeds during stormy rains. High wind speed can destroy the house. Therefore the tree plants help in wind breakage.

However, in the process of juxtaposing the ecosystem services and disservices of homegardens, Ven. Oginni identified the importance of not planting too close to the home building despite the benefit of garden trees as wind breakers protecting the building from high wind velocities. High velocity winds have capacity to damage the buildings of the homes and homegardens come to the rescue from this very important climatic element. Seçkin (2018) argues that the mechanisms of this regulating ecosystem service involve reduction of the wind speed or redirection of the wind movement by the trees. Through these processes the garden trees serve multiple ecosystem services after supplying fruits and vegetative parts like leaves, barks, and roots for medicinal herbal purposes. However, in the view of Mr. Olagunju, the dropping of leaves of trees was identified as an ecosystem disservice considering the need to constantly tidying the home environment to ensure neatness and prevent snakes from hibernating under the leaves. The fear of snake-bite and the consciousness of possible presence of snakes hiding within the green spaces was not an illusion but emanated from experience as snakes have been killed in the compound beforein the home environment of Mr. Olagunju. A similar experience in another homegarden at Omu-Aran was reported that "snakes were plenty". According to the owner of the homegarden when narrating the nocturnal event, the "cashew branches were touching the ground and I put the touch light on it and I saw a very big snake". Adegun (2018) reports that ecosystem disservices are grievous outcomes of green spaces in natural home environments during human-nature interactions. However, these encounters with the nonhuman nature are most likely to be due to lack of fumigation of home environments due to belief in the unsustainable application of chemicals in the homegardens. They also attest to the multispecies ecological urbanism in Yoruba cities where the natural habitats of the nonhumans have been urbanized for human use through incursion into albeit animal territories. Political ecology (Houston et al. 2018) is eloquent about the "imbricated nature of humans and nonhumans" which is "recognised as characteristic of the Anthropocene-multispecies entanglements and becoming-world". It also argues against the "ontological exceptionalism of humans" and its ecological-social certainties that cogenerate the Anthropocene (p. 190).

3.5 Cultural Ecosystem Services of Yoruba Home Greens

Green spaces in Yoruba homes are used for passive recreations. These include strolling, sitting, and love to be in contact with nature of plants. These were cited by many people interviewed across the Yoruba cities. A typical example was Mrs. Faremi O. S., a Yoruba and retired Head-teacher who hails from Okitipupa and had attained the age of 60 years when interviewed in her homegarden at Owo. Active recreation is also enhanced by the gardens. In the process of describing her ecological weed management strategies, she argued that:

> I don't usually use chemicals [*herbicides*] to weed because I don't want any form of chemical effects on my plants. I do the weeding myself […], weeding myself also help me to do some exercise in the morning.

She emphasized that this ecological practice "is more like an exercise, so it helps me [*her*] to have a better health condition. It helps to burn calories in the body and strengthens the heart". Without any gender barrier, majority of the Yoruba people interviewed practise homegardening as an opportunity to exercise their bodies and means of keeping-fit. Mr. Abiola explained that homegardening "[…] is an exercise, you burn out some fat from your system then restore it back from the natural harvest from the garden" and that "it's just a hobby". Mr. Philip Chinedu equally confirmed that the garden was "started like a hobby around the compound" with the involvement of his immediate family members that assisted "in setting up the garden". In this way, homegardening is a form of sociocultural co-creation that enhances social unity, kinship bonding, and filial attachment. This is largely through a process of light hobby that doubles as recreation since the children were helping in the work and playing simultaneously.

Through a cultural perspective and an eco-theological manner, the recognition of a Supreme Being, *Olodumare* (Yoruba name for God) is believed to control and inhabit 'Nature' which they highly reverence in their dealings with the outdoor environments, generally. A Yoruba woman of less than 30 years of age who had Secondary School Education at Ijebu-Ode who planted medicinal herbs and a merchant of herbal concoctions revealed that the herbs were efficacious, asserting that but with "God's decree". Similarly, based on her cultural knowledge of plant seed dispersal mechanisms, she also attributed the growth of the plants to God, that she did not cultivate them but they "spring up from God".

It is amazing to rediscover that cultural ecosystem services are rooted in cultural conveyance from generation to generation within different cultural groups. They establish cultural norms in human relationships with the environment. This can be gleaned from the expressions of Mr. Abiola as follow:

> when I was with my parents, exactly like this structure, they have a waterlogged [*garden*] at the backyard; my daddy will just be there, busy; he's a retired Principal at Lenon Jubilee in Ikare there; you know I have to emulate the idea, the system is within me.

By implication, he claims that this IKS is embedded in his being and value-system as a cultural actor. This suggests that there are intricate connections among human being, traditional socio-spatial norms, pro-environmental consciousness, wellbeing, sociocultural and environmental value-systems through homegardening (Uren et al. 2015). Also, an important cultural ecosystem service in Yoruba ecological urbanism and human ecology is the symbolic indigenous epistemology of proverbial connectedness with plants and nature generally through folklore in their IKS and ways of life (Coker 2018). It is a philosophical method of IKS for achieving wellbeing. Mr. Olukoye explained this with *Iroko* tree (*Milicia excelsa*, formerly *Chlorophora excelsa*, Moraceae family) in the Odugbemi Medicinal Plants Garden. He posited that "*Ati kekere ni a tin pa ekan iroko, ti o ba dagba tan a gba ebo*" (meaning, The roots of Iroko tree are trimmed at a very young age, when it is old it will require sacrifice). The sacrifice referred to is implicitly and explicitly symbolic of the difficulty, rigour, and almost impossibility in trimming the roots or uprooting the Iroko tree without considerable efforts. This is the reason the tree is regarded as a symbol of power, longevity, endurance, and triumph that has capacity to grow for over 500 years. This is with claims that *Iroko* tree has capacity to cure up to 45 diseases including epilepsy, madness, convulsion, impotency, snake-bite, dysentery, and anaemia, among others in addition to being used as anti-witchcraft elsewhere (Ouinsavi et al. 2005). Accordingly, if diseases are also treated at the earliest inception of their risk factors, wellbeing is most likely to be guaranteed.

This ethno-medical approach to wellbeing is not only popular in Yoruba ecological urbanism, it is believed to be more efficacious compared with industrially manufactured drugs. To buttress this ideology, Madam Esther described the significance of her homegarden boastfully that

> you can see I have so many medicinal plants. I rarely go to the hospital. So many times, maybe my children are stooling, they have stomach pain, [...], at times when my children are sick of malaria, I get herbs from the garden, because I grew up to love herbs more than these pharmaceutical drugs. I believe more in these herbs because I am seeing what I'm taking. Those pharmaceutical drugs you don't know what they are made of, so when I take these herbs I feel better and with less stress. I take them, [*it*]is as if I'm taking my normal food, it doesn't look strange to me that I don't know what it was made up. I know the composite of what I'm taking. I take it with my children too.

Her claims reinforce the transferof the ideological belief in the efficacy of ethno-herbal strategies for wellbeing compared with western medicine in Yoruba region. However, parts of the claims raise questions on the proofs of branding IKS as generational ideologies and also the slow pace of acceptance of ethno-medical practices into science. She appears to be highly opinionated that she knows the content of the herbs. Sensory perceptions of sight, taste, smell, and feel can definitely not authenticate the physiochemical contents of herbal medicines in the same way as western medicine beyond textual inscriptions.

Despite this challenge of IKS on the verification of contents ofherbal medication for wellbeing and beyond ideological belief, there appears to be some experiential agreement on certain claims in a number of climes in the region. The narrated experience of Mr. Olabanji is a vivid illustration of such claims. He recalled that he had "always seen" scent leaf "as just a vegetable" until he had ulcer "and the pain persisted even after visiting the hospital". The medical diagnosis of his health condition was correct but he was to undergo surgery. He captured his experiential assessment of IKS of wellbeing in his own words as follows:

> I was diagnosed for operation and I was already thinking of where I'll get the money. Luckily, I met a sister in church who is into herbal drugs and she told me to go [*and*] take [*consume*] scent leaf. I got home and took [*consumed*] the leaves. Before, I only eat once a day because of the pain I had in the stomach as a result of the ulcer but the day I took [*consumed*] the scent leaf, I no longer felt the pain again and since that time, ulcer is a thing of the past. Whenever it occurs again, it's due to my carelessness in feeding. It also works for typhoid [*fever*]. Usually, when you make tea from it and drink it for 3-4 days just to clear your body system.

In addition to testifying to the efficacy of ethnobotanical strategies to achieving wellbeing, his narration validates the role of economic limitations in accessing western practices for wellbeing even by the elite folks in the region when the costs are unaffordably high. Again, his experience reveals the importance of herbal medication as a thriving component of social capital for human wellbeing and the informal mode of its operations in the Yoruba region of Nigeria.

Homegardens are also sources of sociocultural bonding in Yoruba ecological urbanism and a major component of its social capital in many ways. The gardens foster communal cohesiveness and neighbourliness. Owners of homegardens are sources of refuge to neighbours by either giving them voluntary access to the gardens for sourcing medicinal herbs or supplying such needed materials to them as occasion demands especially at critical moments. According to Mr. Olagunju, such a life of giving that produces communal cohesiveness also comes with 'joy of giving'. He explained that possessing homegarden herbs like leaves ofmango and soursop and other items that neighbours request for their wellbeing at any point in time makes him joyful as a giver. Also, fruits are harvested from the garden to entertain guests instead of financial expenditure for procuring industrially produced beverages. Madam Esther cited her own example that her homegarden is "helpful to my neighbours. Sometimes they come, Ma, I'm having this and that challenge, can I get this?" Conversely, she also identified some neighbours as the means of sourcing the plants propagated in her homegarden. In describing the interwoven concepts of social capital, attachment to nature, landscape planning, and sociocultural ecosystem services of homegardens, she affirmed that her homegarden wasplanned to be for some psychological reasons. She affirmed as follows: "I enjoy nature, so when I wake up in the morning and come outside and see these beautiful plants, it reminds me of nature". This form of place attachment to nature exemplifies biophilic relationship of humans with nature. It shows that there are entanglements between human and nonhuman as multispecies urban nature. To arguethat the spaces of homegardens are dynamic to function as social places with communal characters, Madam Esther affirmed that "at times I sit under the mango tree, take some fresh air, discuss with my children and neighbours that do come to visit". Such ambience of passive outdoor recreation doubles as means of adaptation to peaks of thermal discomforts of hot indoors in the tropics. In this regard, the influence of homegardens with grown-up trees in effecting microclimatic quality, amelioration of the harshness of tropical climate, and thermal comfortin urban dwellings cannot be overemphasized. Attachment to nature also has capacity to influence human decisions of where to be and where not to be at specific times and for what purposes. Mr. Olabanji bared his mind on this in his homegarden by drawing example from his dilemma on proposed schedule of activities for the next day of the interview:

> It gives satisfaction that you are under a serene atmosphere. In fact, I have an engagement tomorrow and I'm already disturbed because I want to enjoy the fresh air I have in my compound and that is why I don't like going out.

It was not coincidental in the process of the interview that a neighbour approached the compound of Mr. Olabanjito rest and engage in passive recreation under the mango trees, as he hinted:

> Take your time, he's my neighbour. He does come and rest under my trees, under my mango trees. That's what I'm just telling you now, even when his friends come, everybody will come and stay here. They feel at home here because of the ornamental trees.

In another homegarden scenario, a Yoruba man at the old age of 68 years who holdsa bachelor's degree was interviewed in his homegarden at Omu-Aran in Kwara State. He also attested to the sociocultural ecosystem services provided by trees in homegardens which serve as social spacesof green canopies. He recalled:

> Eight days ago, we did a wedding [*ceremony*] here and we did not need any [*artificial*] canopy […]. We ought to rent three canopies but these trees were used […], and that was where the king sat.

He added that "during heat period, my friends and I do sit on the slab outside, some will say they are leaving here around 12a.m. [*midnight because*] people do say that we have a cool compound". Green canopies provided by trees in homegardens stimulate social ambience for get-togethers in Yoruba cities. Homegardens are thus social spaces that foster community building, friendliness, cohesion, and overall public societal wellbeing and healthy living of individual members. These happen through the enhancement of capacities for building positive emotions like belongingness, communal spirits, and inter-relationships among individual membersof Yoruba society.

Furthermore, homegardening is motivated by interests. This was the claim of ProfessorAkinradeyowho recalled that his homegardening started as an interest in the past. Then, he had always had it in mind that "green environment" was good for him as an artist, foreseeing a futuristic "emotional attachment" to the landscape and the aesthetic satisfaction that he would obtain from it. Environmental ambience and emotional attachment to places through psychological identification with spaces have been correlated with human wellbeing with strong bearings on outdoor leisure behaviour and nature connectedness (Basu et al. 2020; Yuan and Wu 2021). Such place attachment manifests and functions as sense of place, place dependence, affect, identity, satisfaction, and a host of pro-environmental behaviours that enhance wellbeing and quality of life (Ramkissoon 2016; Adedeji and Fadamiro 2018).

However, plant sourcing for propagation presents a challenge to homegardening in the cities. This is either due to lack of access to the desired plants or high cost of procurement. Despite the challenge, love for natural environments and green spaces is a motivational factor for ensuring all barriers to access are removed through sacrificial commitments in cash and kind. An elaborate illustration of this scenario was given by Mr. Olabanji who revealed that he spent heavily in the process of sourcing for the plants in his homegarden. The cost and source of planting materials for cultivation of horticultural gardens are major considerations in achieving the best ecosystem functions (Kramer et al. 2019). The choice of native or exotic plants depends on the overall goal of the owner of homegarden. McKay et al. (2005) argue that developing guidelines for sourcing planting materials is a strategy for avoiding or minimizing mal-adaption to site conditions and this would assist in achieving desirable ecosystem services (Crutsinger 2016) functions.

There are manifestations of understanding of plant metabolisms in the environmental IKS displayed in the case studies of Yoruba people. These metabolic pathways were discussed by Madam Estheras the determinants of planting mix and spacing in her gardening endeavours. She identified the need to apply these ideas to ensure all the homegarden plants are able to adequately access solar energy for photosynthesis. This is to secure propagation survival, efficient growth, and quality yields of medicinal herbs, vegetables, fruits, and other crops. In the application of this indigenous knowledge, plant canopies and vegetative characteristics were considered in selecting plant varieties for propagation. Accordingly, each variety of plant was propagated and their planting location was ensured to achieve adequate reception of rays of sunlight directly. Similarly, IKS are demonstrated in the soil management and nutrient circulation based on soil water content of homegarden. To this end, Madam Esther embarks upon artificial irrigation during dry seasons. She was bold to emphasize that "things like medicinal plants that we use day in and day out, it is easy for us to plant because we have been planting it from generations". The soil management strategies in homegarden IKS she employed also include identification of soil types to determine the required organic and animal waste manures. With her practice of integrated homegardening of flora and fauna, she narrated that she buries little animal dung organic manure around the plants on the already loose grained loamy soil. Loamy soil is already rich in nutrients for planting her vegetables while the sticky clay soil would always require more organic manure for improvement of soil quality. Portions of the garden normally flooded by runoff water during the peak period of rainy season were raised with used automobile tires filled with soil as planting platforms. The practices and applications of indigenous agricultural knowledge have been appraised to support environmental quality and sustainability (Singh et al. 2018). Eagan and Dhandayuthapani (2018) stresses that indigenous knowledge has been used for generations by ancestors in different cultural groups for efficient crop production. Also, the use of IKS in soil management and all other agricultural concerns of homegardens has been proved to be successful even in modern times in many climes, including the whole of Sub-Saharan Africa (George and Christopher 2020; Karki et al. 2021; Mihu et al. 2022; Materechera 2021).

In view of the case studies of ecosystem services of homegardens in Yoruba cities metanarrated in this chapter and the corroboration of the key results with current literature, there is the need for the incorporation of IKS into the frameworks of city planning of home environments in the region. Therefore, the lived experiences of the Yoruba people in their relationship with the nonhuman ecosystems in the cities call for urgent revision of the planning and building laws. The extant planning and building laws have colonial origins and emanated from the cultural lifestyles and worldviews of the Global North. Thus, the current situation is averse to the expectations in urban political ecology where stakeholders should normally be involved in the formulations of ecological policies that affect them through participatory approaches (Shackleton et al. 2019; Sahraoui et al. 2021; Bissonnette et al. 2018). The next Chapter of the book discusses the distinctions between biophilic ideology and sacralisation of natural elements in Yoruba urbanism. This is to examine the distinctions between the Global North and Global South on human relationships with the physical world of ecological elements of urban nature, and how such distinctions could inform the development of evidence-based eco-cultural and ecologically sustainable cities.

References

Adedeji JA, Fadamiro JA (2018) Effect of contiguity of homes and gardens on residents' quality of life in Akure, Nigeria. In: Akinbamijo OB, Okoko EE, Omole FK, Popoola OO (eds) *The just city: poverty, deprivation and alleviation strategies*. Department of Urban and Regional Planning. The Federal University of Technology, Akure, pp 168–180

Adegun OB (2018) When green is grievous: downsides in human-nature interactions in informal urban settlements. J Urban: Int Res Placemaking Urban Sustain 11(3):347–361

Afzal M, Gupta G, Kazmi I, Rahman M, Afzal O, Alam J et al (2012) Anti-inflammatory and analgesic potential of a novel steroidal derivative from Bryophyllum pinnatum. Fitoterapia 83(5):853–858

Agize M, Asfaw Z, Nemomissa S, Gebre T (2022) Ethnobotany of traditional medicinal plants and associated indigenous knowledge in Dawuro zone of southwestern Ethiopia. J Ethnobiol Ethnomed 18(1):1–21

Ajibade B (2006) Anti-bullet charms, lie-detectors and street justice: the Nigerian youth and the ambiguities of self-remaking. In At the international conference, youth and the global south: Dakar, Senegal pp 13–15

Akbar S (2020) *Caesalpinia bonduc* (L.) Roxb. (Fabaceae/Caesalpiniaceae). In: Handbook of 200 medicinal plants. Springer, Cham. pp 483–489. doi:https://doi.org/10.1007/978-3-030-16807-0_51

Ares G, De Saldamando L, Giménez A, Deliza R (2014) Food and wellbeing. Towards a consumer-based approach. *Appetite* 74:61–69

Baker N, Steemers K (2019) Healthy homes: designing with light and air for sustainability and wellbeing. Routledge

Basu M, Hashimoto S, Dasgupta R (2020) The mediating role of place attachment between nature connectedness and human Well-being: perspectives from Japan. Sustain Sci 15(3):849–862

Bissonnette JF, Dupras J, Messier C, Lechowicz M, Dagenais D, Paquette A et al (2018) Moving forward in implementing green infrastructures: stakeholder perceptions of opportunities and obstacles in a major north American metropolitan area. Cities 81:61–70

Bordeianu R (2009) Maximus and ecology: the relevance of Maximus the Confessor's theology of creation for the present ecological crisis. The Downside Review 127(447):103–126

Carrus G, Pirchio S, Mastandrea S (2018) Social-cultural processes and urban affordances for healthy and sustainable food consumption. Front Psychol 9:2407

Casamonti M, Risaliti L, Vanti G, Piazzini V, Bergonzi MC, Bilia AR (2019) Andrographolide loaded in micro-and Nano-formulations: improved bioavailability, target-tissue distribution, and efficacy of the "king of bitters". Engineering 5(1):69–75

Clayton S (2007) Domesticated nature: motivations for gardening and perceptions of environmental impact. J Environ Psychol 27(3):215–224

Clewell AF, Aronson J (2006) Motivations for the restoration of ecosystems. Conserv Biol 20(2):420–428

Cocks M, Shackleton C (2020) Situating biocultural relations in city and townscapes: conclusion and recommendations. In: Urban nature. Routledge, pp 241–268

Coker O (2018) Folklore, indigenous epistemology and legal consciousness: the Yoruba example. Scholarship and Commitment: Essays in Honour of GG DARAH, Chapter 7, pp 49–60

Crutsinger GM (2016) A community genetics perspective: opportunities for the coming decade. New Phytol 210:65–70

Cruz-Piedrahita C, Howe C, de Nazelle A (2020) Public health benefits from urban horticulture in the global north: a scoping review and framework. Glob Transit 2:246–256
Dike IP, Obembe OO, Adebiyi FE (2012) Ethnobotanical survey for potential anti-malarial plants in South-Western Nigeria. J Ethnopharmacol 144(3):618–626
Eagan S, Dhandayuthapani UN (2018) Organic agriculture: Techniques to improve crop production. In: Eco-friendly agro-biological techniques for enhancing crop productivity. Springer, Singapore, pp 1–24
Elufioye TO, Oyedeji AO, Habtemariam S (2022) A review of the traditional uses, phytochemistry and pharmacology of Bryophyllum pinnatum (Lam.)(Crassulaceae). J Biol Act Prod Nat:1–34. https://doi.org/10.1080/22311866.2021.1988706
Eshete MA, Molla EL (2021) Cultural significance of medicinal plants in healing human ailments among Guji semi-pastoralist people, Suro Barguda District, Ethiopia. J Ethnobiol Ethnomed 17(1):1–18
Fernandes JM, Ortiz S, Tavares RPM, Mandova T, Araujo ERD, Andrade AWL et al (2021) Bryophyllum pinnatum markers: CPC isolation, simultaneous quantification by a validated UPLC-DAD method and biological evaluations. J Pharm Biomed Anal 193:113682
Fred-Jaiyesimi A, Ajibesin KK, Tolulope O, Gbemisola O (2015) Ethnobotanical studies of folklore phytocosmetics of south West Nigeria. Pharm Biol 53(3):313–318
George MV, Christopher G (2020) Structure, diversity and utilization of plant species in tribal homegardens of Kerala, India. Agrofor Syst 94(1):297–307
German LA (2004) Ecological praxis and Blackwater ecosystems: a case study from the Brazilian Amazon. Hum Ecol 32(6):653–683
Ghimire R, Joshi N (2020) Complementarity between the homegardening and livestock production systems in Nepal. In: Home gardens for improved food security and livelihoods. Routledge, pp 169–190
Goyal A, Sharma A, Kaur J, Kumari S, Garg M, Sindhu RK et al (2022) Bioactive-based cosmeceuticals: an update on emerging trends. Molecules 27(3):828
Grainger BM (2022) The place of nature in metaphysical religion: a brief history. In: The future of metaphysical religion in America. Springer, Cham, pp 17–41
Hadavi S, Kaplan R, Hunter MCR (2015) Environmental affordances: a practical approach for design of nearby outdoor settings in urban residential areas. Landsc Urban Plan 134:19–32
Herman K, Ciechanowski L, Przegalińska A (2021) Emotional Well-being in urban wilderness: assessing states of calmness and alertness in informal green spaces (IGSs) with muse – portable EEG headband. Sustainability 13(4):2212
Hossain S, Urbi Z, Karuniawati H, Mohiuddin RB, Moh Qrimida A, Allzrag AMM et al (2021) Andrographis paniculata (Burm. f.) Wall. ex Nees: an updated review of phytochemistry, antimicrobial pharmacology, and clinical safety and efficacy. *Life 11*(4):348
Houston D, Hillier J, MacCallum D, Steele W, Byrne J (2018) Make kin, not cities! Multispecies entanglements and 'becoming-world'in planning theory. Plan Theory 17(2):190–212
Idowu EB (1978) African traditional religion: a definition. SCM Press, London
Ingole RD, Thalkari AB, Karwa PN, Zambare KK, Shinde PS (2020) Bryophyllum pinnatum: a magical herb. Res J Pharmacogn Phytochem 12(3):162–167
Islam MM, Shova NA, Rahman T, Anwarul ABM (2019) Crescentia genus of medicinal plants: a review. J Med Plants 7(3):112–116
Jahrl I, Moschitz H, Cavin JS (2021) The role of food gardening in addressing urban sustainability–a new framework for analysing policy approaches. Land Use Policy 108:105564
James K, Wire N, Bradfield JD, Moore S (1979) The holy bible. World Publishing Company
Kang SI (2020) The garden of Eden as an Israelite sacred place. Theol Today 77(1):89–99
Karki KB, Sherchan DP, Panday D, Ghimire R (2021) Soil fertility and nutrient management. In: The soils of Nepal. Springer, Cham, pp 111–126
Kaszás L, Alshaal T, Kovács Z, Koroknai J, Elhawat N, Nagy É et al (2020) Refining high-quality leaf protein and valuable co-products from green biomass of Jerusalem artichoke (Helianthus tuberosus L.) for sustainable protein supply. Biomass Convers Bior:1–16

Kelly JJ (2018) Rogue mystics: the ecology of cosmic consciousness. In: Parsons WB (ed) Being spiritual but not religious. Routledge, pp 181–199

Kramer AT, Crane B, Downing J, Hamrick JL, Havens K, Highland A et al (2019) Sourcing native plants to support ecosystem function in different planting contexts. Restor Ecol 27(3):470–476

Lal R, Bouma J, Brevik E, Dawson L, Field DJ, Glaser B et al (2021) Soils and sustainable development goals of the United Nations: an International Union of Soil Sciences perspective. Geoderma Reg 25:e00398

Latif A, Ashiq K, Qayyum M, Ashiq S, Ali E, Anwer I (2019) Phytochemical and pharmacological profile of the medicinal herb: Bryophyllum Pinnatum. JAPS: J Anim Plant Sci 29(6):1528–1534

Liu R, Grunert KG (2020) Satisfaction with food-related life and beliefs about food health, safety, freshness and taste among the elderly in China: a segmentation analysis. Food Qual Prefer 79:103775

Longo V (2018) The longevity diet:'how to live to 100... longevity has become the new wellness watchword... nutrition is the key' VOGUE. Penguin, London

Lwonga ET, Ngulube P, Stilwell C (2011) Challenges of managing indigenous knowledge with other knowledge systems for agricultural growth in sub-Saharan Africa. Libri 61(3):226–238

Maan AA, Nazir A, Khan MKI, Ahmad T, Zia R, Murid M, Abrar M (2018) The therapeutic properties and applications of Aloe vera : a review. J Herb Med 12:1–10

Materechera SA (2021) Relevance of indigenous knowledge systems in sustaining agriculture under smallholder farming in sub-Saharan Africa. In: Sustainable development in Africa. Springer, Cham, pp 37–55

Matthys B, Vounatsou P, Raso G, Tschannen AB, Becket EG, Gosoniu L et al (2006) Urban farming and malaria risk factors in a medium-sized town in cote d'Ivoire. Am J Trop Med Hyg 75(6):1223–1231

McKay JK, Christian CE, Harrison S, Rice KJ (2005) "How local is local?" a review of practical and conceptual issues in the genetics of restoration. Restor Ecol 13:432–440

Mehta S, Sharma AK, Singh RK (2021) Pharmacological activities and molecular mechanisms of pure and crude extract of Andrographis paniculata: an update. Phytomedicine Plus 100085

Mihu I, Borah D, Tangjang S (2022) Presence of native trees in rural tribal homegardens doesn't necessarily mean conservation: a case study from Mishmi belt of Arunachal Pradesh, India. Ecol Quest 33(1):7–13

Nicholls E, Ely A, Birkin L, Basu P, Goulson D (2020) The contribution of small-scale food production in urban areas to the sustainable development goals: a review and case study. Sustain Sci 15(6):1585–1599

Nogeire-McRae T, Ryan EP, Jablonski BB, Carolan M, Arathi HS, Brown CS et al (2018) The role of urban agriculture in a secure, healthy, and sustainable food system. Bioscience 68(10):748–759

Nolan-Spohn HK (2005) Urban ecology: history and practice (Doctoral dissertation). Boston College, Chestnut Hill

Nwidu LL, Nwafor PA, Vilegas W (2015) The aphrodisiac herb Carpolobia: a biopharmacological and phytochemical review. Pharmacogn Rev 9(18):132–139

Nwonwu FO (2021) 10 Inhibitors of indigenous knowledge systems-led rural village development in South Africa. In: Sirayi M, Kanyane M, Verdini G (eds) Culture and rural–urban revitalisation in South Africa: indigenous knowledge, policies, and planning, p 140

Ogidigo JO, Anosike CA, Joshua PE, Ibeji CU, Ekpo DE, Nwanguma BC, Nwodo OFC (2021) UPLC-PDA-ESI-QTOF-MS/MS fingerprint of purified flavonoid enriched fraction of Bryophyllum pinnatum; antioxidant properties, anticholinesterase activity and in silico studies. Pharm Biol 59(1):444–456

Ojo J (2016) Pesticides use and health in Nigeria. IFE J Sci 18(4):981–991

Òkéwándé OT (2017) A semiotic investigation of philosophical relations between Ifá and Ayò Ọlọ́pọ́n among the Yorùbá people of Nigeria. Nokoko 6:317–346

Olawa BD, Arimih IC, Umeh CS, Arinze NC (2020) The development and validation of a measure of fear of charm within the Nigerian population. Psychol Stud 65:504–512

Omojokun OS, Oboh G, Ademiluyi AO, Oladele JO, Boligon AA (2021) Impact of drying processes on Bryophyllum pinnatum phenolic constituents and its anti-inflammatory and antioxidative activities in human erythrocytes. J Food Biochem 45(3):e13298

O'sullivan F, Laird E, Kelly D, van Geffen J, van Weele M, McNulty H et al (2017) Ambient UVB dose and sun enjoyment are important predictors of vitamin D status in an older population. J Nutr 147(5):858–868

Ouinsavi C, Sokpon N, Bada O (2005) Utilization and traditional strategies of in situ conservation of iroko (Milicia excelsa Welw. CC Berg) in Benin. For Ecol Manag 207(3):341–350

Paniagua-Zambrana NY, Bussmann RW, Romero C (2020) *Bixa orellana* L. BIXACEAE. In: Paniagua-Zambrana N, Bussmann R (eds) Ethnobotany of the Andes. *Ethnobotany of Mountain Regions*. Springer, Cham. https://doi.org/10.1007/978-3-319-77093-2_39-1

Parveen R, Parveen B, Parveen A, Ahmad S (2019) Andrographis paniculata: from traditional to nano drug for cancer therapy. Nanomater Plant Potential:317–345

Pushpakumara G, Sokolow J, Sthapit B, Sujarwo W, Hunter D (2020) Keeping it close to home: home gardens and biodiversity conservation. In: Home gardens for improved food security and livelihoods, 46–77

Raj A, Jhariya MK, Yadav DK, Banerjee A, Meena RS (2019) Soil for sustainable environment and ecosystems management. In: Sustainable agriculture, forest and environmental management. Springer, Singapore, pp 189–221

Ramkissoon H (2016) Place satisfaction, place attachment and quality of life: development of a conceptual framework for Island destinations. Sustainable Island Tourism: Competitiveness and quality of life. Cabi, Wallingford, pp 106–116

Richard EJ, Murugan S, Bethapudi B, Illuri R, Mundkinajeddu D, Velusami CC (2017) Is Andrographis paniculata extract and andrographolide anaphylactic? *Toxicol Rep 4*:431–437

Roe J, McCay L (2021) Restorative cities: urban design for mental health and wellbeing. Bloomsbury Publishing, London

Ruiz-Colmenero M, Bienes R, Eldridge DJ, Marques MJ (2013) Vegetation cover reduces erosion and enhances soil organic carbon in a vineyard in the Central Spain. Catena 104:153–160

Sahraoui Y, Leski CDG, Benot ML, Revers F, Salles D, van Halder I et al (2021) Integrating ecological networks modelling in a participatory approach for assessing impacts of planning scenarios on landscape connectivity. Landsc Urban Plan 209:104039

Salomé M, Arrazat L, Wang J, Dufour A, Dubuisson C, Volatier JL et al (2021) Contrary to ultra-processed foods, the consumption of unprocessed or minimally processed foods is associated with favorable patterns of protein intake, diet quality and lower cardiometabolic risk in French adults (INCA3). Eur J Nutr 60(7):4055–4067

Santiago Fink H (2016) Human-nature for climate action: nature-based solutions for urban sustainability. Sustainability 8(3):254

Saxena S, Pant N, Jain DC, Bhakuni RS (2003) Antimalarial agents from plant sources. Curr Sci:1314–1329

Seçkin NP (2018) Environmental control in architecture by landscape design. A/Z ITU J Fac Archit 15:197–211

Seko J, Bain E, Maponya P (2021) Assessing the impact of indigenous knowledge systems on sustainable agriculture: a case study of the selected communities in the city of Tshwane metropolitan, Gauteng Province, South Africa. In: Venkatramanan V, Shah S, Prasad R (eds) Sustainable bioeconomy. Springer, Singapore. https://doi.org/10.1007/978-981-15-7321-7_9

Shackleton RT, Adriaens T, Brundu G, Dehnen-Schmutz K, Estévez RA, Fried J et al (2019) Stakeholder engagement in the study and management of invasive alien species. J Environ Manag 229:88–101

Sharma P, Sanadhya D (2017) The king of bitters," Andrographis paniculata": a plant with multiple medicinal properties. J Plant Sci Res 33(1):117

Sibeko L, Johns T (2021) Global survey of medicinal plants during lactation and postpartum recovery: evolutionary perspectives and contemporary health implications. J Ethnopharmacol 270:113812

Šiftová J (2021) Shaping the urban homegarden: socio-ecological forces in the management of private green spaces. Land Use Policy 111:105784

Singh P, Singh M, Lone BA, Singh P, Qayoom S, Raina LAS, Kanth RH (2018) Various preferred indigenous agricultural knowledge applications in hill agriculture-a step towards organic agriculture. J Pharmacogn Phytochem SP1:2899–2903

Siriyong T, Phungtammasan S, Jansorn S, Chonsongkram N, Chanwanitsakul S, Subhadhirasakul S, Voravuthikunchai SP (2020) Traditional Thai herbal medicine as an alternative treatment for refractory chronic eczema. Explore 16(4):242–249

Solomon D (2020) Walking the commonplaces of paradise: pastoral precursors in Milton's garden of Eden. The Seventeenth Century 35(6):773–798

Song W, St Clair R, Wang S (2008) Modernization and the sedimentation of cultural space of Harbin: the stratification of material culture. Intercult Commun Studies 17(1):22–37

Stone N (2019) The four rivers that flowed from Eden. In: Studies in Armenian art. Brill, pp 148–172

Swan SH, Colino S (2022) Count down: how our modern world is threatening sperm counts, altering male and female reproductive development, and imperiling the future of the human race. Simon and Schuster, New York

Swanson L, Andruczyk M, Fox LJ, French S (2009) Urban water-quality management: wildlife in the home pond garden. Virginia Cooperative Extension, 426–045

Taiwo O, Xu HX, Lee SF (1999) Antibacterial activities of extracts from Nigerian chewing sticks. Phytother Res 13(8):675–679

Talal ML, Santelmann MV (2020) Vegetation management for urban park visitors: a mixed methods approach in Portland, Oregon. Ecol Appl 30(4):e02079

Teoh ES (2016) Processing of medicinal herbs. In: Medicinal orchids of Asia. Springer, Cham, pp 51–57

Tondi P (2019) The significance of indigenous knowledge systems (IKS) for Africa's socio-cultural and economic development in the dawn of the fourth industrial revolution (4IR). J Gend Inf Dev Africa (JGIDA) *8*(1):235–241

Uren HV, Dzidic PL, Bishop BJ (2015) Exploring social and cultural norms to promote ecologically sensitive residential garden design. Landsc Urban Plan 137:76–84

von Haaren C, Lovett AA, Albert C (2019) Landscape planning and ecosystem services: the sum is more than the parts. In: von Haaren C, Lovett A, Albert C (eds) Landscape planning with ecosystem services, *Landscape series*, vol 24. Springer, Dordrecht. https://doi.org/10.1007/978-94-024-1681-7_1

Wang MY, Ma ZL, He CL, Yuan XY (2022) The antioxidant activities of flavonoids in Jerusalem artichoke (Helianthus tuberosus L.) leaves and their quantitative analysis. Nat Prod Res 36(4):1009–1013

Wilk SR (2014) The scarecrow of Os: the function of antefixes, "Oscilla", and suspended masks in the Roman garden. Classical World:383–392

Wood SA, Tirfessa D, Baudron F (2018) Soil organic matter underlies crop nutritional quality and productivity in smallholder agriculture. Agric Ecosyst Environ 266:100–108

Woznicki SA, Cada P, Wickham J, Schmidt M, Baynes J, Mehaffey M, Neale A (2020) Sediment retention by natural landscapes in the conterminous United States. Sci Total Environ 745:140972

Yakubu FB, Akoun J, Anamayi ES, Duyilemi OP (2008) Mineral composition of selected tree species used for chewing stick in southwestern Nigeria. J Agric For Soc Sci *6*(1)

Yuan KS, Wu TJ (2021) Environmental stressors and Well-being on middle-aged and elderly people: the mediating role of outdoor leisure behaviour and place attachment. Environ Sci Pollut Res:1–10

Zheng M, Zhou Z, Luo Y, Zhao P, Mo J (2019) Global pattern and controls of biological nitrogen fixation under nutrient enrichment: a meta-analysis. Glob Chang Biol 25(9):3018–3030

Chapter 4
Between Biophilia and Sacredness – Global North and South Divide

Abstract The goal in this chapter is to examine the differences (and points of confluence) between the worldviews of the Global North and Global South on the ecology of cities and their ecosystem services. The chapter is hinged on the distinctions between biophilia and spiritualisation of urban nature in the two global regions, respectively. This is in view of modernisation, westernalisation, and globalisation that tend to universalize urban nature. These ideologies framed the hitherto subsistent planning and building regulations in many former colonies in the Global South, including the Yoruba cities that form the South-west of Nigeria. While both biophilia and spiritualisation of nature lead to human wellbeing through apparently similar experiences of nature, the pathways of their mechanisms are different, with some points of confluence on biophilia. The chapter is structured to achieve the following objectives: (1) evaluate biophilia in Yoruba cities to underpin the significance of urban ecosystem services to human wellbeing in ecological urbanism; (2) assess the cultural practises of horticultural gardening in private and public spaces of Yoruba cities; (3) analyze dialogues on spirit ontologies and mysticism as cultural ecosystem services in Yoruba urbanism. Accordingly, the chapter is replete with discussions on biocentric indigenous lifestyles of Yoruba people on strategies of shaping the natural environment for human wellbeing. It showcases biophilia as a major portion of ecological urbanism within cultural context where respondents cited their innate connection with the nonhuman nature as the authority of God. Ecosystem services of horticultural plants in urban settings discovered in the study include security, safety, aesthetic appreciation, medicinal purposes, and mysticism, among others emerge. Large commercial horticultural gardens in the cities are described as healing gardens in view of the therapeutic ecosystem services of the richly colourful plants and aromatic fragrance of their flowers. The chapter concludes with recommendations on means of achieving balance between biophilic goals and spiritualisation of nature in Yoruba ecological urbanism. The next chapter of the book focuses on the formulation of evidence-based design and planning strategies for translating the indigenous knowledge system of Yoruba people on urban ecosystem services into policy tools.

J. A. Adedeji, *Ecological Urbanism of Yoruba Cities in Nigeria*, Cities and Nature, https://doi.org/10.1007/978-3-031-34688-0_4

Keywords Biophilia · Sacralisation of nature · Urban nature · Mysticism · Urban horticulture and agriculture · Human-nonhuman connectedness

4.1 Sacralisation of Urban Nature in Yoruba Cities

Urban nature is perceived by different people, in different climes, in different ways, from one generation to another. For this reason, ecological urbanism in Yoruba cities is hinged on distinctions between secularness and sacredness, biophilic ideologies and mysticism. Also, these are generally applicable to both the Global North and South respectively. While the Global North generally makes no difference between the material world of nature and the non-material celestial system, the Global South generally imbibes a metaphysical sacredness of the material world. Despite the role of globalisation forces in dispersing the dominant metaphysical ideology on environment in the Global South, these distinctions persist in the urban ecologies. The distinctions agree with the definition of urban landscapes as "integrative and visible socio-ecological systems" (Landscape Europe 2015, p. 16) where the ecological systems are essentially natural. The inscription of Osun Grove in Osogbo, Nigeria, by UNESCO as a World Heritage Site makes it suitable for examining how elements of nature in cultural landscapes are spiritualised. Osun Grove is an archetype of urban groves in Yoruba cities. The Grove affords the opportunity of investigating the phenomenology of landscapes in the ecological urbanism of Yoruba cities with field data. This is with a view to facilitating the understanding of how urban landscapes are perceived as precursor to their ecosystem services and how such lessons can inform reconnection with nature in cities. These perceptions are multisensory and deal with how elements of the natural environment and cultural landscapes of the Grove are sacralised in the animistic Yoruba traditional religion to obtain multiple ecosystem services. Plate 4.1 shows the animistic fence of Osun Grove and its ecological setting. Figures 4.1 and 4.2 shows how River Osun is deified for traditional worship. Figure 4.3 shows the entrace to *Ifa* forest as a section of the Grove. Figure 4.4 shows the *Iya Mopo* shrine of the Grove. The grove is used for Osun worship and adherents troop into its yard based on their belief-system in Osun goddess as shown in Fig. 4.5. The grove also serves urban ecotourism purpose including majority of its sections and shrines.

These Figures show the extent to which elements of nature have been mixed with animistic human creations of sculptural elements to produce the Grove as a cultural landscape. The metaphysical perceptions of nature as evidenced in the spiritualisation of its elements in the Osun Grove are proofs of innate connection of humans with the natural world. These perceptions are forms of 'inner' 'sacred' biophilia expressed through traditional religious worship. However, there is another form of biophilia that is 'non-sacred', 'non-religious', and therefore Western in its rationalism but is equally expressed in non-Western contexts like among the Yoruba people.

Fig. 4.1 The animistic fence of Osun Grove Osogbo in Nigeria and its ecological setting. Source: Photograph by author

4.2 Biophilic Rationalism in Yoruba Cities

Ecological urbanism is inherently rooted in harmonius relationship of humans with the natural world. This relationship is entirely bio-centric throughout human history and civilisation. It captures human nature as a biological organism living, working, and recreating in the nonhuman nature of the environment through 'non-sacred' biophilia. In human history, co-living with plants, animals, air, natural light, water, weather conditions, and the whole range of terrestrial, arboreal, and aquatic ecosystems is paramount to human survival. Satisfying direct experience of nature has also been central to human and environmental sustainability and has capacity to determine longevity of life (Kellert and Calabrese 2015). Professor Gabriel has deep knowledge of ecology. He was interviewed in his herbal garden at Warri. His responses showcase the benefits of wellbeing through experience of nature and its contributions to human wellness and longevity of life. He exclaimed, "Oh yes, if you go a natural way you can look forward to clocking a good 100 years". The jettisoning of nature and natural processes in cities and what the ecosystem services of nature could contribute to wellness is the bane of contemporary urbanism. Eventhough nature is wanted by all humans because of the intrinsic biophilic relationship with nature, urban dwellers are often allienated from nature because of non-ecological urbanism. In view of his own agerasia which he connected with 'natural' lifestyle and oneness with nature, Professor Gabriel added his perspective that:

> Being at home with nature is what everyone wants. All the hustles and bustles in the city, stress and so on. If somebody continues like that, by the time the person gets to around sixty [...] you look older. I have so many people around me, even in the church here, you know, those who are in their sixty, 'am older than they but they can't, even when they go out I see them bending.

Fig. 4.2 River Osun showing spiritualisation of elements of nature for traditional worship ecosystem service in Osogbo, Nigeria. Source: Photograph by author

This therapeutic significance of biophilic urbanism cannot be substituted with any affordance of techno-urbanism in view of its sustainability and naturalistic mechanisms. Therapeutic spaces mirror healing gardens where emotional attachments to nature are paramount. These multisensory attachments to nature produce multiple ecosystem services in urban green spaces, including biological and horticultural gardens.

For instance, Mr. Adebisi Abioye, a Yoruba man from Irele in Ondo State of Nigeria, at the age of 49 years and holding a diploma certificate was interviewed at his MTK Biological Garden, Akure. The garden contains security, medicinal, and beautification plants. He noted that the practice of gardens and gardening started with the biblical Adam and Eve making gardening "the profession that first came to existence before other professions started". While navigating the garden, he pointed to a plant and emphasized: "This one is a type of hibiscus flower known as Variegated hibiscus. It's herbal and beautification plant [...] used for treating pile when used with other plants and also has other herbal uses". His assertion implies that non-native ornamental plants could double for ethnobiological use in Yoruba region. To drive home his point of dual use of plants for both beautification and consumption for wellbeing, he introduced another plant as he explained: "This one is called Fir,

Fig. 4.3 The entrace to *Ifa* forest as a section of Osun Grove Osogbo in Nigeria. Source: Photograph by author

Fig. 4.4 *Iya Mopo* shrine, Osun Grove, Osogbo, Nigeria. Source: Photograph by author

Fig. 4.5 Adherents of Osun goddess worship troop into the Osun yard inside the Osun Grove Osogbo in Nigeria based on their belief-system in the goddess. Source: Photograph by author

commonly called the Christmas tree. It's both medicinal and for beautification, for treating malaria". Furthermore, some ornamental plants serve dual purposes of beautification and security. Security plants around the fence make entry hard for intruders to "enter into the house", he explained as he pointed to an example: "This tree is called Pongamia (*Pongamia pinnata*, F. Fabaceae). When you move closer to the tree you will notice some thorns on it. These thorns are poisonous but from afar it is a beautiful plant". Another plant in the garden, but mainly for beautification, is Asura (*Asura cervicalis*, F. Erebidae). To link ornamental gardening with socioeconomic statuses, Mr. Abioye added that the garden was mostly patronized by the elites, the political class, and expatriates like Indians in the city. This implies a strong association between imbibing western worldviews and practices of ornamental gardening. He explained that the volume of patronage of the garden is also dependent on planting season.

These explanations are evidences that ornamental homegardening in Yoruba region is not meant for only environmental aesthetics but also serve non-aesthetic purposes, including human wellbeing. Other ecosystem services of ornamental homegardening in the region are security, safety, and medicinal purposes. Accordingly, Mr. Abioye added that "green bush is [*a*]security and medicinal plant that poke the skin and it would be difficult to remove". He identified its use for treating chronic pile "when mixed with other medicinal plants and consumed raw". Plants like Bush rose (*Rosa hybrid*, F. Rosaceae) and Bachelor's button (*Centaurea cyanus*, F. Asteraceae) are exotic ornamental plants in the garden that serve security purposes. It is also common to find anti-snake plants in the Yoruba region that provide safety ecosystem service by scaring snakes away from home environments througth their fragrance and scents.

In another sense, the MTK Biological Garden, Akure, was also described as a healing garden by Mr. Yinka. According to him:

> When the rainy season comes, you will be smelling aromatic fragrances. It is scientific, some of my friends that always come here, they don't usually get sick, even my children when they are on holiday, they will come here and play here, they too don't get sick. If a heart-diseased person even comes here, with the help of the aroma, they will be relaxed with the beautiful things they see […]. When it comes to your health, staying inside the air conditioner is very dangerous. When it comes to this place, your health will improve even your blood pressure will not rise.

This means of wellness through connection with natural landscape is one of the biophilic advantages of nature. Heerwagen (2009) links health and wellbeing to biophilia through urban landscapes that have capacity to produce restoration by therapeutic mechanism. According to Lumber et al. (2017), nature connectedness through compassion, meaning making, emotion, and aesthetic appreciation is a pathway in biophilia. Urban gardens and green spaces are unique environments that foster such biophilic pathway that enhances everyday experience of nature through contact with the natural world (Lin et al. 2018). For similar reason, Mr. Abioye argued that there is connection between wellness and horticultural homegardening, and that the two are intricately and significantly connected. He underscored his argument on the significant contribution of urban horticultural gardening to the stock of green spaces and experience of nature in Yoruba cities as follows:

> Everybody thinks horticulture is based on flower [*plants*] alone. No, we have some fruits section. Different types, apple, citrus, mango, guava, different species of mango, different types of citrus. So that section alone covers lands, if you have about two acres just for nurseries. I have different types of citrus. During rainy season now, I'll travel to Benin[*and*] Benue to get my fruits from there and I'll go to Ministry of Agriculture at Ibadan to get my fruits".

Horticultural gardens in Yoruba cities are sources of plant procurement for propagation for all purposes, including homegardening, landscaping of public spaces, and urban agriculture.

Furthermore, the biological life processes in biophilia are most compatible with the ideals of human-nonhuman nature connections and easily satisfy the needs for daily ecosystem services. Ven. Oginni bared this fact when describing his homegarden at Owo. He explained that:

> As a matter of fact, in the evening you just go round the whole place. By the time you come back you are fresh, apart from the fact that you want to take something from the farm [*garden*]. The breeze that comes out of it also helps the body. It takes care of the body, so I enjoy that very seriously. I enjoy to see nature because God creates it for our own use. Sincerely, if one is sick and you just move round the compound like this, before you know it you are well again, because God has created it that way.

The attribution of divine providence to God in nature discourse and biophilia is expatiated in the paragraph that follows. However, it suffices to iterate here that there is a complete acknowledgement of a divine order in most environmental processes of life in Yoruba cities, with the indigenous knowledge systems explored in all matters of being.

Accordingly, exposure to the natural environment has been confirmed to be of high value in enhancing human wellbeing and overall quality of life (Adedeji and Fadamiro 2018b; Hall and Knuth 2019; Garrett et al. 2019; Giannico et al. 2021). Though biophilic ideology has its roots in Western science, its manifestations are universal but through different mechanisms. It posits that humans have a tight afinity with nonhuman nature that cannot be broken. This intricate relationship reflects the interest and care for other forms of life by all humans in their daily environmental experiences. An earlier wrong understanding of urbanism as isolation of humans from the natural world paved way to designing nature out of cities. Cities were interpreted as dominance of built forms over natural elements with a notion of human dominance over nature instead of seeing nature as co-dwellers in cities for care and nurture. The design and building of cities with an ideology of conquest of nature and natural environment (Blackbourn 2011) have taken their tolls on urban dwellers who are made to pay for human actions taken against nature (Studer 1998). Little wonder that Engels (1964, p. 183) decries:

> Let us not, however, flatter ourselves overmuch on account of our human victories over nature. For each such victory nature takes its revenge on us.

A critique of the philosophies of Francis Engels and Karl Marx on ecology (Löwy 2017, p. 10) establishes that human societies and nature are in "disastrous metabolic rift" even in the twenty-first century due to the "destructive consequences of capitalist expansion of the environment"I. This ecological critique of capitalism attests to the reality of ecological disasters that are anthropogenic. However, with the advent of landscape urbanism in which cities are re-interpreted and understood in terms of the landscape and not buildings, a new hope of biophilia resurged. It is a triumph of the landscape over buildings. In this resurgence, green infrastructure and their ecosystem services became a central discourse with greater attention to nature-based solutions (NBS) (Scott et al. 2016) compared with the machine-age environmental crises in cities. While this has a global reference, the Global North which has been at the center of industrialisation also bears much of these concerns. With less industrialisation and therefore lesser effect of the machine age on the Global South, the urban environmental crises due to anti-nature approaches have been less until the Anthropocene. This relative impact of anthropogenic activities has been agravated to higher revenge of nature and vulnerability of urban dwellers in the Global South due to low coping strategies consequent upon fragile socioeconomic capabilities (Spring 2020; Zigale 2021). The poor and vulnerable groups thus suffer more revenge of nature but their greater biophilic association with nature due to their IKS provides a succor (Santos et al. 2021; Kingsley et al. 2013; Anguelovski and Roberts 2011). Therefore, Global South cities continue to enjoy biophilic relationship with urban nature. This is in addition to the bulk of the sociocultural ecosystem services peculiar to them due to the metaphysical worldviews and notions of nature. In the process, biodiversity conservations are achieved through mythological strategies.

Consequently, biophilia, "the innate tendency to focus on life and life-like processes" (Wilson 1984, p. I), is an essential component of urbanism in the Global South, including Yoruba cities. It deals with environment-behaviour interaction and is dominant in human psychological coping with stress through ecological affordances (Araújo et al. 2019). Biophilia developed with biophilia hypothesis that "proclaims a human dependence on nature that extends far beyond the simple issues of material and physical sustenance to encompass as well the human craving for aesthetic, intellectual, cognitive, and even spiritual meaning and satisfaction" (Kellert and Wilson 1993, p. 20). The varieties of life and life-processes are implicit and explicit on biodiversities of life and clearly connect biophilia with ecosystem services of nature in general and urban nature in particular. Within the categories of urban ecosystem services, biophilic services have been identified as biophilic ecosystem services (el-Baghdadi and Desha 2017).

Despite the universality of biophilia, there appears to be a great disconnect in the Global North from its spiritual component that is inherent in the worldviews of the Global South (de Villiers and Moolla 2018). This disconnection is evident in nature's skepticistic ideology about the spirituality of humans through connections with the elements of nature – plants, animals, water, air, light (Simpson 2022; Sayem 2018). Conversely, few indigenous people in the Global North and most in the Global South are ardent practitioners of nature-based faiths (Preston and Baimel 2021; Simpson 2022). Spiritual nature-based faith is often a Western approach to escaping religion and coding it as a non-scientific discourse. Whether it is accepted, pretended, manipulated, or rejected, it is evident that biophilia exposes the innate connection of all humans with some form of spirituality or another, to which nature is central, but in different shades. Frank Llyod Wright was apt to conclude that "Nature should be spelled with a capital 'N', not because Nature is God but because all that we can learn of God we will learn from the body of God, which we call Nature" (Spirn 1997, p. 249). Considering himself as deeply religious, he introduced the idea of organic architecture in which the built form is subsumed in an environmental naturality. When asked for a substantiation of his religious claim: "Do you go to any specific church?" In response, Wright submitted, "My church [pause], I put a capital 'N' on Nature and go there" (Spirn 1997, p. 250). Wright's form of religiosity through connection with nature is essentially biophilic and Western, where clear distinction is made between religion and science (Wood 2002). In making this distinction, sharp contrasts are drawn between the myths of religion and the laws of science (Farnham and Kellert 2002). Wright's religiosity is distinguishable from cultural biophilic connection with nature that is spiritual as largely found in the Global South (Kellert and Farnham 2002). In describing this Frank Lloyd Wright's organic utopia of architecture in connection with the natural environment, which are both akin to the home and its garden in the paradigm of an ecological urbanism, it is arguable that:

> Besides serving definite aims [*the building*] would express, through its own specific geometry, the essential order of the universe ..... A building was to take possession of its surroundings in order to enhance the beauty of its location, just as lake, a bluff, or a tree enhance the appearance of a landscape. In Wright's utopia, the beauty of buildings would be a permanent object lesson of the joy of communion with nature (cited in Bauman and O'Brien 2020, p.101)

The question is, in what ways do biophilia manifest itself in the cultural ecosystem services of urban green infrastructure in the ecological urbanism of Yoruba cities? This broad question could be approached from two perspectives: biophilic spirituality in the Global North that is more like that of Frank Lloyd Wright, and 'worship'/deification of nature in the Global South. Illustrations would be drawn for both perspectives from some of the interview responses in the study.

The reflection of Professor Jaye Akinradeyo, professor of art, on the environmental aesthetics of his homegarden at Ogbomoso provides some insights into biophilic ecosystem services of homegardening in the ecological urbanism of Yoruba people. According to him:

> It's an emotional thing to satisfy each time I come here. Infact, it doesn't make me stay out for too long now. I thank the environment and I find it too difficult to stay in other environment[*s*]. I always long to come home because of the greenness of the environment. I believe there is a hidden open aesthetics around here. When you wake up, that is why not that I can't do interlocking [*concrete*] but why [sic] everything white, white, white, they are not natural. We are created in a natural environment. So, the green is […] look at my hen here, I'm not able to buy, but few time I'm able to buy them small corn. They substitute it with the green. Imagine if it were interlocking, then you start sweeping.
>
> It is good to leave our environment green, like yesterday, when I came back, I wanted to eat vegetables. I cannot ask them by eight o'clock to go to the market to get it.

Six themes emerge from this response – emotions of landscapes, personification of the nonhuman, place attachment, aesthetics of nature, residential landscape design, and affordances of homegardens. Emotional satisfaction from views of natural landscape of homegardens retells the story of how best to create home environments. Human emotions of excitements and sensations relay feelings of oneness with nature from which most city dwellers have been alienated due to the artificiality of the urban design. Such positive emotions engender joy, peace of mind, euphoria, affection, tranquility, coolness of temper, kindness, cheerfulness, blissfulness, inner strengths, fascination and satisfactions with life. They are products of spaces that are exciting, interesting, exhilarating, arousing, stimulating, sensational, pleasing and beautiful, as shown in the arousing-pleasant continuum of Russell and Lanius (1984) model of affective quality of places. With places implying spaces with character, "the model proposes that emotional reactions to environments can be described by their relative position on unpleasant-pleasant and arousing-not arousing continua" (Adedeji et al. 2020, p. 298). In the pleasant-not arousing continuum, the model includes place affective characters like nice, pleasant, serene, restful, peaceful, calm, and tranquil as positive emotions. These spaces of positive emotions are antidotes against negative emotions like sorrow, sadness, anger, hot temper, and depression, among others. These feelings correspond to the Russell and Lanius' (1984) model of affective quality of places in the not arousing-unpleasant continuum of negative emotions like un-stimulating, dreary, dull, boring, inactive, idle, monotonous, and lazy. The arousing-unpleasant continuum of the model consists of negative emotions like displeasing, unpleasant, repulsive, dissatisfying, uncomfortable, tense, panicky, forceful, intense, frenzied, and hectic.

Professor Akinradeyo reinforced these biophilic emotions of the landscapes of his homegarden by arguing that:

> Even the environment here, people have been coming to take pictures. Apart from the edible things, the environment is also inviting. I want to send this message to the architects, they should encourage green environment. When you have interlocking [*concrete*] around you because you don't want to cut grass. You know, I attended a conference, where we were told that it [*interlocking*] aids cancer.

In distinguishing between biophilic ornamental landscaping and homegardening to achieve aesthetics, Professor Akinradeyo was of the opinion that it is not rewarding enough to embark on ornamental gardening of homes compared with edible or medicinal plants. To him, pursuit of aesthetics should not be limited to planting ornamental flowering plants but could be achieved with gastronomic plants for double purpose. In his words, he said:

> I prefer to use pepper as flower [*-ing plants*]. They are good flower [*-ing plants*]. Okro, they are flowering [*plants*] while plantains are hibiscus. But if you plant flower [*-ing plants*], it's not ediblebut put okro there, you distance them the same way, then you see the flowers and they are very beautiful. These vegetables, if you put one here and another one there, even when you are not eating it [...]. See, is this not beautiful enough? [...] But when you see hibiscus, it's not useful, so why planting it? Have you seen any flowers [*ornamental plants*] around here? [...] Vegetables are my flowers [*ornamental plants*].

In this scenario, aesthetic biophilia doubles for consumptive purposes for provisioning ecosystem services. The reason is that ornamental homegardening is not part of the IKS of Yoruba people in their care for the home environment (Adedeji and Fadamiro 2018b), unlike largely in the Global North where rigorous attention is paid to ornarmental urban home landscaping constituted by lawns and unedible plants (Taylor and Lovell 2014). Incidentally, the reproductive systems of the gastronomic plants that enhance cultural identity and constitute traditional ways of living (Taylor and Lovell 2015) also have flowering mechanisms that are desirable and serve aesthetic functions for positive human emotional satisfactions and affections.

The personification of the nonhuman in Professor Akinradeyo's intuitive perception of nature when he said "I thank the environment", attests to an anthropomorphic frame in this form of biophilia that could be distinguished as 'cultural biophilia'. From this anthropomorphic construction of the natural environment emerges the mother Earth. This is a view of nature as a friend to co-dwell with and not a foe to be tamed (Keune et al. 2022). The ecosystem services of homegardens as major components of urban green infrastructure call for an appreciation from humans who benefit these services. It is an act of thanksgiving to the mother Earth (Matholeni et al. 2020) almost portraying a humanistic filial love in this biophilic relationship. It accounts for the environmental stewardship of humans that comes along with the care, protection, and sustainable use of urban ecosystems. It is an act of environmental consciousness that has gone through a thorough process of observation, understanding, acceptation, appreciation, appropriation, and commendation. Human biophilic connectedness with the earth requires reflective observation of its temporal and material realities, and thoughthful understanding of its life processes. These result in acceptation of its ecosystem service offers, grateful appreciation of its life-giving offers, sustainable appropriation of its ecosystem services, and proactive commendation of its roles in human wellbeing.

There is an expression of place attachment in this biophilic environmentalism. It presents a parallelism with the narrative of Mr. Akanbi a Yoruba man at the age of 50 years with a bachelor's degree when interviewed in his homegarden at Akure. He affirmed the following about his homegarden that:

> [...*It*] has a very wonderful experience I must say. I can say it's a natural habitat for me. In the evening, I go there, I relax, I move closer to the nature because it has so many advantages. When I come back from [...], I go to the place, relax, do so many things, till the ground, remove some weeds from the ground, then engage in wetting and so many things.

The assertion that the homegarden is a human habitat reinforces the co-habitation of multispecies nature in terrestrial ecosystems of ecological cities. Attachment to places with requisite affordances like relaxation is a part of usual opportunities for passive recreation that produces sense of place. The meanings people ascribe to places ultimately determine their levels of attachment to such places. For Professor Akinradeyo, it is a sort of refuge brought about by greenness of his homegarden. This meaning produced visual and emotional aesthetics of nature in his being. Such aesthetics of nature simply implies the desirability of the beauty that natural environments and their intrinsic processes display. A deep philosophical critique of the concept of beauty in nature environmental aesthetics is outside the scope of this book. Glenn Parsons (2008) provides an excellent analysis of this subject in his book, *Aesthetics and Nature*, arguing that the concept of beauty conveys feelings of love, affection, and moral regards. The complexity of the subject of beauty and how it applies to nature remains a continuing intellectual dialogue in environmental psychology. The debate lies in how beauty should be measured and its great variability of meanings from one person to another. The Cartesian revolution marked a distinguishing point in the beauty debate by identifying and separating nature from mind. In this argument, nature is defined as "what is out there" and mind as "what is in here" (Lothian 1999, p. 195). With this distinction, place attachment to natural environments is a product of connection between the mind and nature. Accordingly, Mr. Akanbi affirmed that his homegarden "has so many benefits. One of the benefits is that it makes me move closer to the nature". Such power of attraction to nature materializes from place identity to produce sense of place and its emergent place attachment. It is a manifestation of the centripetal force in biophilia that draws human to nonhuman nature conciously or unconsciously through the inner being.

The emerging quest is how this biophilic attachment to nature informs residential landscape design that incorporates nature and its affordances of homegardens in the ecological milieu of Yoruba cities. Professor Akinradeyo conflates the planning of home environments with ecological considerations like air circulation, microclimate, security, aesthetics, and greeneries. While hard surfaces like concrete should be minimised, green spaces in the form of gardens and green yards should be maximised. The gardens and yards could be located at advantageous positions depending on the goal of home owners. These could be sited in the front, back, or sides of the home building. A similar view was expressed by Mr. Adekomi Joseph who retired as a lecturer from a Federal College of Agriculture in one of the Yoruba cities. To him, scenery of the environment offers 'enjoyment' that makes nature paramount for home planning because "the shade of the plants" affords "fresh air". The landscape planning includes fencing which should be designed to maximize security, defensible space affordance, visual aesthetics and air flow. While challenging and discouraging the idea of tall and closed-view fencing of homes in landscape planning, he reiterated:

> So, why are we disturbing ourselves? You've denied yourself fresh air. So, it [*home fence*] should be something minimal. You should even allow people to see the beauty of your house. It's another form of security. We don't pray for any bad thing to happen. If like you have come in now, if there is any negative intension that you have here, people there are looking at us. The moment it turns violent like this, they'll say what is going on there. So, when it's not a prison yard, it's only a prison yard that will have that.

Residential fencing is an important modern component of landscape planning of home environments in the new and sprawling suburban sectors of Yoruba cities. This is in view of increasing urban insecurity consequent upon widening socioeconomic inequalities (Adedeji et al. 2016). Residential landscape planning with Western biophilic ornamental plants is also gaining ground in Yoruba cities due to colonialism, civilisation, cultural transfusion and globalisation. This is evident in ornamental homegardening enhanced by horticultural practices and merchandise in the ecological urbanism of the cities. The next section of this chapter focuses on these practices to gain insight into their relationships with indigenous homegardening in the cities.

4.3 Horticultural Practices and Urban Agriculture as Eco-city Components of Yoruba Cities

Indigenous homegardening of medicinal plants has high precedence over ornamental homegardening in Yoruba ecological urbanism as discovered so far in this book. However, a line of distinction occurs in the case of public places like hotels, government offices, educational institutions, and public spaces produced from Western ideologies of cityness and urban aesthetics. In view of these, ornamental horticulture is still at low level, with majority of the plants being exotic. Mr. Abiola, a Yoruba man in his early thirties holds a polytechnic diploma, was the manager of Lucado Horticultural Garden in Akure where he was interviewed. He explained:

> This place is dominated by plants [*ornamental plants*] which are being used primarily to beautify homes and surroundings and there are various kinds here namely; variegated hibiscus, ixora, yellow bush, araucaria spp., and so on[...]. Plants are good because they beautify homes and apart from that, plants help in human respiration, that is; they emit oxygen, O_2, which human being breath in while humans emit carbon dioxide, CO_2, which the plants need for growth.

Variegated hibiscus (*Hibiscus rosa-sinensis*, F. Malvaceae), ixora (*Ixora coccinea*, F. Rubiaceae), yellow bush (*Lupinus arboreus*, F. Fabaceae), and araucaria spp. (*Araucaria heterophylla*, F. Araucariaceae) are from exotic sources with araucaria as the most espensive of all. They are the common ornamental plants in the new sectors of Yoruba cities in the region, including government offices and school premises. The submissions of Mr. Abiola explain the biophilic significance of ornamental homegardening and the general pursuit of environmental aesthetics as its goal. It bothers on why a multispecies urbanism is not only significant but a compulsory requirement for human existence since oxygen signifies life (Decker and Van Holde 2010). The metabolism process that makes life of humans and many other fauna exist and survive has oxygen as its base in the mechanism. This has not only been confirmed in science but evident in real life health situations. According to Briehl (2015), “In the absence of oxygen human life is measured in minutes” (p. 124). Therefore, in view of this significant role, ornamental gardening is a significant ‘modern’ component ofYoruba ecological urbanism. However, Mr. Abiola decried that “people place less value on plants”. This may not be unconnected with the notion that ornamental plants are not consumable and therefore do not offer provisional ecosystem servicesbut only supply cultural aesthetic value, along with regulating and supporting services.

The Lucado Garden contributes to the urban ecological system ofAkure city. With its advantaged location near a stream, the garden also benefits from the stream as a source of water for its irrigation system. Mr. Abiola confirmed that the garden also serves as fauna habitat and thereby “contributes greatly in retaining the ecosystem.” Urban ecosystemsof green infrastructure in Yoruba cities also contribute to the hospitality industry in the region. Alli Bolaji Francis, a married Yoruba man and a bachelor’s degree holder, was the horticulturist of Halmark Suites, a popular hotel at Ota in Ogun State of Nigeriawhere he was interviewed. One of the ornamental plants in the hotel landscapewas green ficus (*Ficus benjamina*, F. Moraceae), locally referred to as fence flower or weeping fig which he described as scentless but “no matter how it is trimmed it still brings out the beauty”. Others are murraya (*Murraya paniculata*, F. Rutaceae), which is medicinal and “has scents when it’s around seven or seven-thirty pm [*when*] it’s fragrance starts getting released. That’s its beauty and it smells so beatifully” and bougainvillea (*Bougainvillea species*, F. Nyctaginaceae) which [*produces*] flowers all the time except only 3 months of January to March. Crown-of-thorns also called Christ thorn (*Euphorbia milii*, F. Euphorbiaceae), Irish coconut (*Cocos nucifera*, F. Arecaceae), Royal palm (*Roystonea regia*, F. Arecaceae), and many varieties of palms embellish the hotel grounds. Explaining that the ornamental palms were many but not consumable, he emphasized that “It is just a palm [...]. Its major usefulness is for aesthetics” and “we have about seventeen or eighteen different types of palm [...] we have sycher palm at the front there”. The palms include Traveller’s palm (*Ravenala madagascariensis*, F. Arecaceae), Cycad palm (*Cycas revoluta*, F. Cycadaceae), King palm (*Archontophoenix alexandrae*, F. Arecaceae), Royal palm (*Roystonea regia*, F. Arecaceae), and Queens palm (*Syagrus romanzoffiana*, F. Arecaceae), among others. Canna lily (*Canna indica*, F. Cannaceae), Thuja

(*Thuja occidentalis*, F. Cupressaceae), Agave (*Agave americana*, F. Asparagaceae), Croton (*Codiaeum variegatum*, F. Euphorbiaceae), were planted "just to beautify the compound"mostly for their colourful flowers, thorns and vegetative systems.

Furthermore, horticultural practices are sometimes holistic in Yoruba cities. Beyond seeking aesthetic satisfaction, the Yoruba people combine the planting of varieties of horticultural items. In particular, they sometimes practise homegardening that combines the cultivation of medicinal plants, vegetables, and food crops. These are common sights in the new areas but most especially in the suburban sprawl of Yoruba cities. The planting of crops for provisioning ecosystem services in homegardens presents a blurring of boundaries between homegardening and urban agriculture. Crop like cocoyam (*Xanthosoma saggitifolium*, F. Araceae) is planted for both provisioning and medicinal services. Other crops like varieties of yam, Dioscorea spp. (water, white, bitter, and yellow yam), cassava (*Manihot esculenta*, F. Euphorbiaceae), and maize (*Zea mays*, F. Poaceae) are planted for the consumption of the household. Furthermore, naturally-irrigated vegetable gardening is also done along urban waterscapes during the dry season. All these also contribute supporting and regulating ecosystem services to the urban milieu of Yoruba region. They suggest similarities between gardening for food and medicine, but both contribute to the ecosystem services of urban green infrastructure.

4.4 Oneness with Nature Through Recreation in Yoruba Cities

Formal recreation settings constitute a major group of green infrastructure that contribute significant ways to the oneness of city dwellers, both indigenous and non-indigenous, with nature in Yoruba region. The recreation spaces create opportunities for close interaction not only among humans but importantly between humans and non-humans, including space and 'behavioural' sharings, exchanges, and overall 'appreciation' of other members of the multispecies community. Lekki Conservation Centre Lagos (Fig. 4.6), is one of those recreation spaces with these kinds of multi-species interaction as shown in the Figures that follows. Figure 4.7 shows the tree house of the Centre. Tourists who are agile climb the tree to reach the tree house as an active recreation. It not only mimics the climbing habitsof the monkeys and other faunas in the park but also weaves together an ecological symbolism encapsulated in human-nonhuman connections. Figure 4.8 shows how humans and monkeys use the same space in a very relaxed multispeciesecologicalenvironment. Figure 4.9 shows a 'socialization' scene in the monkeys' family of the Centre. Figure 4.10 shows the 410 m canopy walk through nature at the Centre which is the longest in Africa.

The picturesque of the Lekki Conservation Centre (LCC) are evidences that ecological systems of Yoruba cities are imbued with numerous ecosystem services and major opportunities to explore nature through urban ecotourism. Urban ecotourism at the LCC affords protection of wildlife, biodiversity of nature zone, and

Fig. 4.6 Lekki Conservation Centre, Administrative Block, Lagos, Nigeria. Source: Photograph by author

environmental quality. These opportunities are components of the panacea that urban ecotourism provides beyond utopia (Das and Chatterjee 2015). In the course of exploration, humans can understand natural processes and form pro-environmental behaviours that could enhance environmental sustainability and wildlife conservation. Pro-environmental behaviours of urban ecotourists formed through exploration of natural scenic spots and associated embodied emotion and connectedness with nature are sure pathways to achieving the sustainable development goals of the United Nations and Agenda 2030 (Cao et al. 2022; Esfandiar et al. 2022; Hunt and Harbor 2019). In the Yoruba region, there are aspects of urban pro-environmental behaviours that are ecologically based on spirit ontologies and mysticism.

4.5 Ecological Spirit Ontologies and Mysticism in Yoruba Cities

Spiritualisation of nature is a major component of ecological urbanism of Yoruba cities. The contextual identity of the cities in this regard is part of a broader scholarship. This scholarship distinguishes between the Global North and South in how humans approach, identify, understand, and relate with urban

Fig. 4.7 Lekki Conservation Centre Tree House, Lagos, Nigeria. Source: Photograph by author

nature (Cocks and Shackleton 2020; Shackleton and Cocks 2020). While cultural biophilia and sacredness are clearly embedded in the biocultural identities of the indigenous people in the Global South, less of these are acknowledged in the Global North except those having attachment to non-modern indigenous ways of life (Shackleton and Cocks 2020). The spiritualisation of nature through mystification of ecological elements – plants, animals, water, air, light – delineates the Yoruba cities as clearly belonging to the psych of the Global South on nature beyond geographical classification (Dobbs et al. 2021). The lucidity of this assertion in complicity with religious practices and afflitiations is described in this section.

Alhaji Tiamiyu Ayinde was a Moslem and an Islamic leader but he also practices medical herbalism as a profession (Herbalist -*Babalawo*). Herbalism is not only becoming popular and civilized as alternative medicine, its traditional roots subsist. *Alhaji* Tiamiyu Ayinde was interviewed in his medicinal homegarden in Oyo. He explained the complementarities among herbal practices, spirit ontologies, and mysticism. In discussing the lucrativeness of his herbal endeavours and how it is determinable by these complementarities, he submitted as follows:

Fig. 4.8 Lekki Conservation Centre, human and nonhuman nature occupying the same space at the Visitors' Centre, Lagos, Nigeria. Source: Photograph by author

Fig. 4.9 Lekki Conservation Centre Lagos, Nigeria, showing a 'socialization' scene in the monkeys'ecosystem of the Centre. Source: Photograph by author

Fig. 4.10 Lekki Conservation Centre Lagos, showing the 410 m longest canopy walk in Africa. Source: Photograph by author

> You know buying things at the herbal market (*Lekuleja*) without knowledge is not good. That is why they sell fake items to most people without knowledge. The practice of herbal medicine is knowing what you want to buy and knowing how to use them. That is why most people would prepare charms and it wouldn't be effective. There are some plants you have to say some things (*incantations*) before plucking them. There are some you have to spray some things on the ground before touching them.

These mechanisms reflect the fetishness of herbalism in cultural ecosystem services of green infrastructure of Yoruba cities. Herbal markets in the Yoruba cities are public markets where common and specialised flora and fauna ingredients for metaphysical herbalism and trado-medicinal concoctions are sold. They are located at the traditional core sectors of the cities in the Kings' Markets and are important elements in the cultural worldviews of the people for wellbeing. The preparation of charms for both offensive and defensive strategies are also cultural ecosystem services of nature in Yoruba cities. They are manifestations of the biocultural identity of Yoruba people in particular and characteristic of the Global South. In sourcing these juju and amulet ingredients, an intersection between materialism and spiritualism is engaged such that certain spirits associated with specific flora and fauna are invoked. Such invocation into the spirits world are sometimes believed to be for the apperance of the flora or fauna or for such item to be discoverable in the environment or efficacious for the intended use after sourcing. Sometimes, availabilities and accessibilities of such materials are reliant on temporalities and perceptive modes of fetish discernments. These are forms of traditional religion to which westernly-minded Yoruba people would not suscribe to, especially some Christian and Muslim sects. Those who practise either Christianity or Islam with these indigenous religious beliefs are usually labelled as having multiple allegiance and hypocrisy. This view was expressed by an Education Officer of Osun Grove Osogbo, Mrs. Aduke Tawa, in a study as she emphasized that:

> Some people are hypocrites. You see them go to church and still come here and worship. You see some in hijab, hijab sisters, they are here for spiritual needs, not for sightseeing. They hide their religion, they still practise traditional religion. Even they come here with their hijab. Some, they park their cars and they went to the river to have their spiritual baths. They consult. Some are hypocrites. They hide their religion, practice Islam. Some believe purely in traditional [*religion*] (Adedeji 2020, p. 57).

In relation to these indigenous religious practices, during the interview granted by *Alhaji* Tiamiyu Ayinde, he pointed at *Asunwon* or *Ajarere* (*Casia alata*) plant in his garden. This medicinal plant is believed to be effective for the treatment of human fertility-related problems of wellbeing. He also pointed at another medicinal Croton plant in his homegarden during the interview. He called the Yoruba name of the plant *Ajeofole* or *Eyeobale,* (*Croton zambesicus,* F. Euphorbiaceae). He explained that:

> It is used for rebuking evil spirits. If mixed with some ingredients, it is used to break the bondage of deadlock of witches. The witches' birds can't land on it. If they do, that means their power would fail.

This explains the metaphysical connection that Yoruba people make between the physical ethnobiological world and the spirit world to achieve wellbeing through ecological elements of nature. The witch spirits are believed to 'turn to' birds, a fauna, and believed can be incapacitated by floral spirit believed to be associated with *Croton zambesicus*. This metaphysical belief-system is based on a notion that witches are wicked humans that have capacity to fly and therefore would require a spiritual metamorphosis to turn to flying birds in their nefarious nocturnal activities. The interest in this chapter is not about the details of these metaphysical translational process in spiritism. However, it suffices to note that Yoruba people believe in the realms of the celestial (*Orun*) and the terrestrial (*Aye*) separated and connected by an aerial space where physical and metaphysical birds fly from one point to another (Ayinla and Ayodeji 2022; Balogun 2017). This description is a similitude of a heavenly divinity and earthly ecosystem (Bartholomew and Bartholomew 2012; Hart 2006).

How these intricate layers of operations are embedded in environmental cognition and spiritualisation of nature are inscribed on the city neighbourhood environments in the form of shrines dedicated to certain gods and godesses. These practices have remained despite the forces of urbanisation, civilisation, and modernisation. For instance, Lagos, both metropolitan and sub-urban, has retained these cultural ecological identities as a type of cultural ecosystem service to date. Even though through 'pretended' democratic urban governance, urban developments have led to desacralisation of the groves' foregrounds of Lagos shrines, they are purposely conserved as developments are designed to include them. Therefore, they are still in existence at roundabouts and under urban tropical trees at Surulere (formerly *Abule Esu/Elegba*), Lagos Island, Ojuelegba, Ojota, Egbeda, Idimu, Agege and Ikotun. The multi-nuclei urbanisation of Lagos makes these communities to have separate shrines at the entrance of each since indigenous Yoruba people believe that the shrines are sources of protection for the inhabitants. During an interview conducted at Enu-Owa Lagos, some facts emerged about the ontological connections between ecological nature and cultural practices in Yoruba cities.

The *Enu Owa* Square grove is the ritual centre of Old Lagos. Old Lagos, or Lagos Island, is the traditional urban core of Lagos that is characteristic of Yoruba cosmological city planning. Its shrines function in the cult power relations of the Lagos chieftaincy institution. Enu Owa is located in Isale Eko community and it is very central to any discourse on Lagos, being as old as Lagos itself. Isale Eko is an ancestral and indigenous area in Lagos with many traditional cultures related to the Lagos kingship institution. In relation to the kingship institution and how it is ecologically inscribed on Enu Owa square and Isale Eko in general, an interviewee, Mr. Mahmud Shodeinde, a married 64 years old Yoruba and indigene of Isale Eko submitted that:

> Majority of the houses in Isale Eko are ancestral houses full of cultural and traditional collections. [...] Isale Eko is a place where the palace of Oba of Lagos is situated and majority of the traditional white cap chiefs have their palaces in Isale Eko, either Akarigbere, Abagbon, Ogalade or the Idejos. The Idejos traditional white cap chiefs are the land owners. The Akarigbere migrated from Benin and settled in Lagos as far back as 1630 and the Abagbons guide and protect Isale Eko community while the Ogalades are incharge of the spiritual and traditional activities.

Thus, the traditional and cultural ecosystem services of Enu Owa are unique and many. The Akoko tree at Enu Owa is used for the coronation of kings and chiefs in Lagos as in all Yoruba traditional kingship institutions. Enu Owa houses Oduyan shrine, Olusogbo enclave and Oju Olobun tree, with adhent practices of Yoruba traditional cult religions. The very old Oju Olobun tree and Akoko tree are ecological elements that define this city space as an urban grove. Their contiguities with the traditional market at Enu-Owa speak volume of the intricate connection between socioeconomic and ecological value-systems in Yoruba urbanism. For this reason, these eco-cultural sites are distinguished by traditional sacredness from profane spaces of the city for everyday use. The centrality of these urban shrines are evident in the carrying out of every traditional activities in Lagosat Oju Olobun. With the only traditional market in Lagos located at Enu Owa and combined with the cultural ecological sites, make Enu Owa the traditional spiritual headquarters, ancestral home, and heritage of Lagos. It is an indigenous area with heritage artifacts that have been in existence for over 500 years. A significant neighbourhood at Enu Owa is called Idumoyinbo where the European explorers first visited in 1704 during the reign of Oba (King) Akinsemoyin. These antecedents position Isale Eko and Enu Owa in particular as a "traditional, social, political, economic, [*and*] educational" hotspot of Lagos. The economic prowess is premised upon commercial activities in general and provisioning ecosystem service of fishing in particular. In view of the aquatic ecosystem of the adjacent Lagos Lagoon, the predominant ecosystem service in Enu Owa and Isale Eko in general is fishing.

In his narrative of the evolution of Lagos to unpack the factors of the city's settlement pattern, Mr. Mahmud Shodeinde noted a cultural connection between Enu Owa in Lagos and Enu Owa in Ile-Ife and cultural morphology of Yoruba cities in general. According to Mr. Mahmud Shodeinde, "[…] the first settlers in Lagos, the Aworis, they stayed at Iddọ but people from Benin moved down to this area. So, Enu Owa is a very reserved area and if you go to Ile-Ife, we have Enu Owa [*there*] too" where kings and chieftaincy installations take place with *Akoko* leaves. Just like the Enu Owa at Ile-Ife, the Enu-Owa in Lagos hosts the Ogun Shrine, Odiyon Shrine, and Enu Owa Shrine which are components of Yoruba traditional religion's pantheon. The Ogun Shrine doubles for worship and coronation of Eletu Odibo family chieftaincy title. The Enu Owa Mosque and King's Church are also sacred spaces in this city core and is characteristic of Yoruba city cores.

How these eco-cultural complexities of Global South cities are interwoven with ecological urbanism has been described by Spirn (2011, p. 614) as follows:

> Human survival depends upon adapting ourselves and our settlements in life sustaining ways, designing places that reflect the interconnections of air, earth, water, life, and culture, that help us feel and understand these connections, places that are functional, sustainable, meaningful, and artful (Spirn 1998:26). Ecological urbanism aims to advance this goal.

Idanre is a Yoruba city in Ondo State where these thoughts on ecological urbanism are explicit. The city is home to the popular Idanre Hills (*Oke'danre*). The hills exemplify an eco-cultural urbanism of elements of human and non-human nature in time and space. As the first settlement of the city and its archaeological milieu, the hills document the ancestral origin of the city and how it has grown from cradle in precolonial era to the present status of urban ecotourism site. The ecological combination of hills, waterbody (thunder water – *Omi Aopara*), vegetation, and cultural antecedents of containing the first settlement, Agooogun footprint, ancestral graves, shrines, and palatial building attests to its uniqueness as a sacred landscape. In view of this cultural sacredness, the hills have been deified for ancestral worship and traditional religious practices celebrated as public events during the fifth month (May) of every year. The hilly urban landscape is believed to be inhabited by some spirits responsible for the protection of the dwellers during war times in the past. Accordingly, Idanre Hills clearly fit the description of Langhorst (2014) as an urban ecological site of

> picturesque traditions […], representations and aesthetications of ecological processes […*that are*], central to current ideas of 'landscape urbanism' and 'ecological urbanism' […], 'right to urban nature', 'aesthetics of (ecological) performance' vs. '(ecological) performance of aesthetics (p. 1110)

According to this description, the Hills are perfect representations of "ongoing negotiation and interaction of human and non-human processes" (Langhorst 2014, p. 1112) in distinguishing between ecological biophilia and sacredness through aesthetic performance.

The ecological aesthetic performance of the Hills was investigated through onsite interviews. The Awise of Idanre Kingdom was a high chief and one of the key custodians of the Idanre cultural heritage. He intersects traditional religion, culture, history, environmental aesthetics, and urban ecology in describing the importance of Idanre Hills. According to him:

> If you climbed the hills there [*the*] other time, this is where they call Ife Oke (a flat top). It has been in existence before our forefathers. There is a ritual that is carried out once in a year on the hill. The worshipers wrap white robes during the festival. There are other chiefs like Pa Lawrence, Solo[*mon*], Gbenin, they are the chiefs in charge of the festival every fifth month of every year – Olosun Festival.

He attributed the ecological uniqueness of the Hills to God – "God created the hills with some mighty things that even the mighty cannot confront [*when*] they enter the hill" and gave a narration on the mystic water feature of the Hills. "There is a lake of water at the top of the hill called Ahun River [...]. If anyone has problem and come to pray on that water before use, the problem of the person will be solved". This mystic ecology is attributed to the spirit of the hills believed to be responsible for wellbeing by the indigenous Idanre people. According to Awise, Idanre Hills contribute progress to the lives ofthe city dwellers who usually experience advancement at every festival. They believe that prosperity and fortunes accrue to the dwellers of the city through traditional spiritual interventions connected with the hills. Conversely, ill lucks manifested through setback in life endeavours are believed to be curable through ritual sacrifices on the hills. It is also a common belief in Idanre that the religious festival attracts communal peace and tranquility whereas absconding from its observance could result into environmental calamities like strange windy storms destroying houses and fatal auto-accidents.

Oladayo Benard was a tourist at the Idanre Hills when interviewed there. In his attribution of the hills to an eco-theological providence, he acknowledged the hills as the creation of God that has popularized the city as an ecotourism destination of global significance and an outdoor learning site about nature. In a similar perspective, Adeola, a Christian resident of the city submitted that the hills "serve as a tourism [*site*] for people from far and near. It makes Idanre more prestigious than other towns that don't have this natural endowment". In the view of Baba Ijebu who resides in the city, Idanre is rich in environmental uniqueness "where people come to see the wonders of God". The city has a legacy of marvelous admirations through the Idanre Hills as tangible and intangible heritages of the Yoruba generally and Idanre in particular. Baba Ijebu exclaimed – "Can you see that wonderful tree on top of the hill with flourishing leaves?" and "If someone comes here to pray to be rich, the person will definitely be rich". These affirmations of cultural ecosystem services are rooted in a traditional eco-theology that "makes Idanre more known by outsiders and also increases our belief that a Supreme Being do [sic] exist" as opined by Adefoye Ifeoluwa, a Christian lady who was resident in Idanre.

Adedeji and Fadamiro (2018a) underscore the centrality of monarchical institutions to landscape evolution in traditional-modern Yoruba cities. Therefore, it is pertinent to recall relevant historiographical responses on Idanre Hills from an interview granted by the monarch of Idanre, Owa of Idanre Kingdom, Oba Fredrick

Adegunle Aroloye IV, and reported in the Punch Newspaper of 29th April 2017 (Dada 2017). According to this first class Yoruba monarch,

> Our former name was Ife-Oke. When our children were being kidnapped, we sought protection and left that area for Utaja. There was a man among us, he was a hunter, and he said he discovered a place on the hilltop that would accommodate our population at that time and get us away from the threat of kidnappers. That was how we moved up there and we were there for 800 years before we started coming down again (Dada 2017).

In view of his historical sketch of the evolution of Idanre and the role of the Hills in the sustainable growth of the city, the Hills and the associated grove constitute both tangible and intangible heritages for cultural ecosystem services. Beyond these, the Hills document the British colonial administration in Yoruba land. Oba Fredrick Adegunle explained this British colonial complicity in the location of the city as follows:

> In 1894, Governor Carter from the colonial office, came up there to sign a treaty with my grandfather, Arubiefin the First, saying her Majesty, the Queen of England, would give them protection and that they should not follow any other foreign country. So he was enthroned in 1919. In 1926, my father wrote a letter to the Lieutenant Governor in Enugu. Then, Nigeria had been divided into northern and southern protectorates and the southern protectorate was being administered from Enugu. So it was Enugu that my father sent the letter to and it was debated in the legislative council and passed that we were free to move down the hilltop. It was then the Surveyor-General was asked to survey the new town for us. They made two places for us called Ilu-tuntun (Dada 2017).

These historical accounts are clear antecedents that Idanre Hills have been central to the evolution, continuity, growth, and sustainability of Idanre Kingdom. The accounts also show the hegemonial role of colonialism in framing colonial urbanity of Idanre Kingdom and how this interlaces with landscape heritage preservation. The sacralization and eventual deification of the hills obviously arose from the protection that the settlers enjoyed during their period of dwelling there. The contiguity of sacred groves to cities is characteristic of Yoruba cities that always exist with groves dedicated to certain gods and goddesses of the Yoruba pantheon. This mirrors a central goal of ecological urbanism through cultural ecosystem services where human nature lives with, and not over, non-human nature in a symbiotic relationship for the sustainability of the duos. The abiotic elements of the Idanre Hills including rock, water, air, light, and vegetation that supplied the material and metaphysical needs of the settlers for over 800 years in an ecological milieu. The complicity of colonialism in the development of Idanre is related to political ecology. It led to the relationship between nature and society and the environmental outcomes of the power relations. These concerns reflect components of ecological urbanism, including land occupation, public space and habitability, mobility and services, urban complexity, green areas and biodiversity, urban metabolism, and social cohesion (Cormenzana et al. 2014). In another sense, the narrative is an eloquent proof of how ecology has "undergone a series of conceptual permutations" as an investigative apparatus to analyzing the city, resulting in the evolution of genres of urban ecologies and their traslations into ecological urbanism (Gandy 2015, p. 150).

The anthropological antecedents of Idanre Hills highlighted in the foregoing discourse are instrumental to its submission for enlisting as a World Heritage Site with the UNESCO in 2007. In view of this, based on the genuineness of the site as "one of the most awesome and beautiful natural landscapes […] with inselbergs of about 3,000ft above sea level" UNESCO describes the site as having:

> diverse and variegated eco-systems of flora and fauna, […] very important bio-physical and land form features whose interaction with the physical features created an enduring cultural landscape within the setting. […] The festivals provide occasions for reconciliation of the low-landers and their natural environments as well as reenactments of historical episodes in local Idanre history and its wider Yoruba ideology, mythology and confederacy. […] Apart from the festivals, which make the site a living tradition, the flora and fauna of the hills arc also unique. There is a special specie of tailless animal called Hydrax that lives on the rocks […] and there are special monkeys spotted near Orosun hills. Added to its beauty which fires human curiosity is the fact that the entire people of Idanre lived on these boulders for almost a millennium (UNESCO, n.d. – https://whc.unesco.org/en/tentativelists/5169/).

Modern concrete steps numbering 660 have been introduced to the hills for ease of climbimg for ecotourists that patronize this urban nucleus site. The exhibition of Yoruba mythology in Idanre Hills is common in the ecological urbanism of all Yoruba cities of all scales, sizes, and urban systems. For instance, in the case of Enu Owa Lagos earlier described, how 'the urban' is conceptualized from ecological standpoint of mystic nature in cities is revealed in the presence of Oju Olobun shrine, Odiyan shrine, Osugbo shrine, and Eletu Odibo palace, and Eletu Odibo being the custodian of the Odiyan shrine. The last Eletu Odibo was late Chief Tajudeen Gbadesere Eletu and the vacancy had not been filled during the course of the data collection for this book. The person of Eletu Odibo is the principal and head of the King Makers of Lagos, who operates through the shrines. This confirms Lagos as a mystic cultural urban place. The accompanying traditional rites for the coronation of Oba of Lagos and white-cap chiefs are carried out at Enu Owa using the Akoko tree. Other ritual activities include an annual ritual prayer performed for urban sustainability, progress and success by all traditionists in Lagos at Oju Olobun shrine containing the Oju Olobun tree.

During the interview granted by Mr. Mahmud Shodeinde in Lagos Island, he described the significance of the venerated Oju Olobun tree at Enu-Owa. According to him "the Oju Olobun tree has been there since the time of Eletu Alijosun, that'll be around 1908" and is presently protected by blockwall. This attests to the power of indigenous religions to conserve ecological resources for a very long time even in an urban space like Lagos Island. It reinforces the arguments in ecotheology that indigenous religions are supportive of ecological conservation. There are huge debates about the centrality of spirituality to Afrocentric environmental ethics and their impetus to establish Afro-American religions (Manzini 2019; Ssebunya et al. 2019). Urban ecological realities and their climate change scenarios show that reinvigoration of natural green spaces in cities are required through reorientation on the value of trees. Despite the politics of religio-climate change in environmentalism (Bauman and O'Brien 2020), the power of religion to achieve this in Africa has been a major focus in African ecotheology (Torabi and Noori 2019; Green and Haron 2020; Conradie 2020).

Accordingly, Lagos urbanity does not foreclose mystic ecological practices. Oju Olosa grove shrine (for the deity of lagoons) is located at the Wharf alongside the Lagos lagoon. Yoruba traditional belief system reflects in the common identity of planning settlements around groves as also evident in Lagos. The practice of Ifa oracle divination, *Oro*, *Oree*, *Egungun*, *Elegba*, *Igumuko*, *Opa*, *Osugbo* and *Gelede* festivals by the *Aworis* are evidences of a mythological settlement pattern in time and space. These were grossly desacralised based on high pressures on land use due to urbanisation. Still, they have been sustained in Yoruba ecological urbanism.

Conversely, a central theme in Western ideology is the distinction between what constitutes science and myth (Scott 1996), thereby rejecting mythical notions of ecological urbanism from the cannon of science. Science adheres to procedures in investigating naturally occurring phenomenon in the physical world and is generally implied to be denial of supernatural accounts. However, the biotic and abiotic components of ecological urbanism appear to be in agreement with the definition of myths as "ethnocentric views of the actions of humans, living and non-living objects, natural and super-natural, and metaphysics of the factual and fictional" (Adedeji 2020, p. 54). With the inclusion of biotic and abiotic realities in both scientific and mythic notions of ecological urbanism, a curiosity arises on the fate of myth for inclusion into scientific paradgims. The argument of Nasr (2006, p. 207, 708) that "A science of nature based upon power and dominance over nature rather than the contemplation of its ontological and symbolic reality reigns supreme as the only legitimate form of knowledge and is almost deified and certainly absolutized". This is where the challenges in scientific canonisation lie. With the idea of organic nature, capitalist nature, and techno-nature (Escobar 1999) enshrined in scientific categorization of ecological components in the natural world, myth is implicitly included as a part of cannons of science. With the inclusion of 'constructed' nature in organic notion, myth is part of its animistic and totemistic ontologies. In an ecological sense, including its urbanistic conceptualisations, constructed nature is configured with interwoven relationships among the natural world of human society, animals, plants, spirits, gods, goddesses, and their positions in Indigenous Knowledge Systems and biocultural diversities. This mode of living with nature is contradictory to the idea of 'dominated' nature in capitalist concepts overloaded with regulation, simplification, and dominance over nature through governmentalisation, commodification, production and conquest of ecological components of cities. If the manipulation of these components in artificial nature, genetic engineering, simulations, and hybridisations (Escobar 1999, p. 9) is recognized as science, why not mystic urban nature which are "representational forces given gendered personalities for human perception" (Mowatt 2018, p. 7).

In an interview at one of the Oduduwa Peoples' Congress (OPC) Centre, Lagos, granted by Mr. Kareem Ahmed, a traditionalist, Moslem, and head of the Centre, it was surprising to discover that Lagos is still hosting these mythological practices despite westernisation. The OPC Centre is a cultural landscape that has worship shrines for *Esu* and *Ogun* demarcated with tropical tree. Ogundipe (2018) wrote extensively in his book, *Esu Elegbara: Chance, uncertainty in Yoruba mythology*, to argue against a widespread belief equating and depicting *Esu* with the notorious devil/satan in Judeo-Christian theology. In his account, *Esu* is a Yoruba god and not the fallen Lucifer, and both have differing characteristics in the opposite directions, he argues. In the introduction to the book, Olupona explains the role of *Esu* as a messenger god, and believed by the Yoruba indigenous tradition that *Esu* is the closest deity to *Orunmila*, the Yoruba god of divination. With similar perspective and emphasis, Mr. Kareem Ahmed told the story and the reasons behind the traditional worship at the Centre. Affirming that the personality called devil has been in existence before humans, he submitted that the worship of *Esu* is a tradition traceable to the Yoruba progenitor Oduduwa. Both *Ogun* and *Esu* are co-worshiped at the Center with the claim that both deities co-operate mythologically and people who worship them together are co-protected by the duos, they believe. He explained that, *Esu* is historically believed to be a troublesome god through a complex mode of operation, a part of the spirit world, which has to be placated by those who hold the traditional religious belief. According to Mr. Kareem Ahmed, when both gods are appeased, they guide and defend their worshippers, which he explained to be an ageless experience in Yoruba traditional religious system. He explained further that, often than not, a favourite worship of either *Esu* or *Ogun* only is believed to breed dangerous rivalry of gods for the worshippers and assumed to lead to a 'fight of two elephants' where the worshippers suffer.

In these complexities with claims of "more deeper and wider" metaphysical activities, Mr. Kareem Ahmed identified Orunmila as the super god regulating, coordinating and controlling all the activities of the gods in Yoruba pantheon. These beliefs and practices are not limited to the followers of traditional religion. Some people who claim to be in the practice of Christianity and Islam are found to engage in traditional religion saying "Orunmila is likening [sic] unto God that we call in the Church". According to Mr. Kareem Ahmed, these set of people equally hold that traditional religion "is what has been in existence" for long and cannot be forgotten. He was of the opinion that "not that you become a Christian and will say you don't do such again, then decided to burn them" since the traditional religions have been inherited from one generation to another. These accounts establish the multiple religious practices in Yoruba cities that sustain traditional ecosystems despite westernisation. This is much similar to what Adedeji (2020) termed "hypocrisy, multiple allegiance and religious unification" (p. 57). The practices are not delineated nor limited to socioeconomic groups and locations. There are many deities in the Yoruba cities and there was a claim by Mr. Kareem Ahmed that "there are some children that are spiritual" in traditional ways. The worship of these deities is accompanied with incantations, sacrifices, and traditional religious decorum. These are dictated by *Orunmila* such that the conducts of the worshipers are guided by the gods, as the indigenous Yoruba people believe. To emphasize this, Mr. Ahmed recalled that *Orunmila* could instruct him and other traditionalists "to worship Ogun with a fowl" which they should carry out.

Animal sacrifices in Yoruba traditional worship is a reminder on the interconnectedness of the physical and metaphysical, the flora and the fauna, the human and non-human components of ecological urbanism and their eco-cultural complexities. The unseen gods are metaphysically located at shrines marked with floral trees where fauna sacrifices are offered for the appeasement of the deities as believed by the Yoruba traditional worshippers. The traditional worshippers believe that the gods are akin to humans in gastronomic demands and dainties including drinks. In the explanation of Mr. Ahmed on what could be termed as 'the dainties of the deities', he argued that:

> Now we worship it (that is, the *Esu* and *Ogun*) the way we worship it, we offer it its own food (its own soft drink for instance as we humans like softdrink) [sic]. If it is to go to a higher level, you will be demanded to offer a goat to worship it regardless of the cost. You could be mandated to offer a cow to worship *Ogun* as sacrifice, it is cow you would offer. The most important sacrifice *Ogun* eats is dog. Dog is its best food, and fowl. [...] If you want to worship *Elegba* [*god*], a ram and chicken would be killed; they will make pounded yam with *egusi* [*melon soup*] they will cut plenty meat in a basin; cut the chicken like ten [*pieces*]; then they will take it to *Elegba* shrine with the egusi and pounded yam. That is what we do use to worship *Elegba*.

These narratives are shrouded in mysteries and raise many questions. It is not clear how the sacrifices are claimed to be accepted by the gods though the narratives describe the procedures for the worships in Yoruba ecological mythology. At the Oduduwa Peoples' Congress centre, items of sacrifice were still physically present at the shrines during the interview. This occurrence is typical of all shrines and points of sacrifices in Yoruba cities. It is a common sight at road junctions to find most elements of sacrifices offered to the gods intact for days while some are carted away by mentally-challenged people, especially if such sacrifices are gastronomically dainty. It is also a common taboo in Yoruba cities that sacrifices offered to the gods and deities should not be eaten by humans to avoid negative consequences believed that the gods could inflict. The practices also result in unsightly outlook of shrines bearing heterogeneous combinations of sacrifice item. These are offered in-situ and in particular order accompanied with chanting of incantations meant to invoke the spirits of the gods. The blood splashes of slaughtered sacrifice animals and poured drinks can characterize the shrines. In this way, the shrines are not only unsightly but may also emit odours of decaying elements which are usually decomposable organic items. These appear to be sociocultural ecosystem disservices in Yoruba cities.

According to Mr. Ahmed Lawal, Yoruba people believe that Ogun god is a mobile element of nature and becomes the head of any settlement reached. For this reason, Ogun shrines are scattered all over Yoruba cities, with appellations like *Ogun Aro*, *Ogun Onire* at Ire-Ekiti. The god is believed by the traditional Yoruba people to be present in every Yoruba Kingdom and has two symbols, fire and bloody cloth, which they attribute with ferocious anger. Mr. Ahmedcredited these practices to mind-body dualities in indigenous religions and boasted of their efficacies for his own personal wellbeing through mind perception of current or future events. Through eloquent boasting, he affirmed the multiplicity of religious practices of his members as follows:

> When herbalists are performing their ritual, they do make mention of *Eleduwa* (God). It is God that created everything. Everyone here, for example myself, I am a dedicated Muslim, this guy here is a Christian. Everyone here do [sic] attend a place of worship. All these things here are because of our traditions.

This multiple religious allegiance attests to the persistence of indigenous Yoruba religions in the cities despite the advent and practices of Christianity and Islam to very large extents. However, many Yoruba people have unalloyed allegiance to only one religion. Mr. Ahmed equally recalled the involvement of his grandmother in the worship of *Esu* (also referred to as *Elegba* god) at Oju-Elegba (literally meaning the shrine of *Elegba*) in Lagos Island. He grouped *Sango* among the gods and referred to them as kings, and gave example of *Sango* (god of thunder) as a protector of the worshipers. Within this indigenous belief-system, while attending to its mythological taboo relating to ecology, Mr. Ahmedemphasized the possibility of the deities displaying anger by attacking biological elements of nature like trees:

> There are some things in this world that God created. For example, if there is a tree here and the king of thunder realize he has not strike for a while, he will attack the tree, the tree will die immediately. In Yorubaland, if *Sango* (god of thunder) kill any type of tree, such tree cannot be used as firewood [...]. The tree Sango kill cannot be used as firewood in Yoruba tradition. But if we are in a place like this, that is city, we don't know if it is *Sango* that kill a tree, we then use it to cook. In such house, there will be a problem without us knowing the cause; this is because the tree *Sango* kill must not be used as firewood. If *Sango* kill a tree in the forest, any Yoruba man who gets there will know that it was *Sango* that kill the tree. So, the tree will decay there without anyone using it.

This narrative of spirit ontologies and mysticism on ecological taboo can only be an ecosystem disservice. Conversely, the same *Sango* is ontologically believed to enhance ecology by attributing rainfall to be part of the activities of the god when sent on revenge mission during the dry season when it does not normally rain in Yoruba region. Other forms of these ontologies were discussed by an Herbalist (*Babalawo*) at Oyo, *Alhaji* Tiamiyu Ayinde, who inherited herbal practices. According to him, certain plants offer mystic protection of building from rainstorm damages when incorporated into the construction of the building. An example of such 'mystic' plant in his home medicinal garden was Sodom apple or Dead Sea apple (*Calotropis procera*, F. Asclepiadaceae) called *Ajandi* or *Bomubomu* in Yoruba language. This plant has provisioning ecosystem services, its dried and powdered roots being used in the ethnomedical treatment of certain diseases like epilepsy, sores, asthma, among others. It is also used for such mystic construction purpose in Yoruba traditional ecological urbanism. In the words of *Alhaji* Tiamiyu Ayinde,

> The stems and branches of the plant are being cut in half inch or a foot and used in roofing of houses; they are placed in the four corners of the houses in the roofing process before the roofing sheets are being added. It's for prevention from wind and storms; no matter how strong the wind is, the roof would not be affected. The storm can, maybe carry another roof and place it on the roof of the house (where *Bomubomu* is used) but the storm can't do anything to the roof.

This claim is evidence that building construction process in Yoruba traditional urbanism could be laden with cultural complexities. Such mystic spiritualisation of ecological elements of nature extends to everyday life in the type of plants around home environments. *Alhaji* Tiamiyu Ayinde pointed to another plant calledEmpress candle plant or Candle bush (*Senna alata*, F. Fabaceae *Asunwon* or *Ajarere* in

Yoruba). According to him, this male and female fertility plant used in the treatment of athritis, eczema, and many other diseases, also has mystic uses:

> It is used for rebuking evil spirits; if mixed with some things, it is used to break the bondage and deadlocks of witches; the witches' bird can't land on it, if they do, that means their power would fail.

These indigenous knowledge practices are passed orally from one generation to another. They are crafted by a mystic notion of nature that all physical elements of the material world have underlying metaphysical backgrounds. In this knowledge system, physical events are attributed to spirits and gods both implicitly and explicitly.

The fetishes of intermingling ecologies of spirits and ecologies of the material world make non-traditional Yoruba people like the Western world to regard spiritualisation of nature as magical, paganism, and ritualistic (Ryan 2019). These ecologies are generally characteristic of the Global South and Afrocentric myths in particular (Zimmerer and Bassett 2003). They are a bundle of complex interplay of spiritism (believing in the existence of, and communicating with the invisible spirit world), animism (believing that all the physical elements of the material world have or are associated with certain spirits), ancestral worship, and divination. These complexities differentiate religious spiritualisation of nature from biophilia which is essentially Western and generally Global North dominant worldview. Different Yoruba people cultivate natural materials like plants but use them in different ways to explore different ecosystem services. These differences come to fore in the dialogue with *Alhaji* Tiamiyu Ayinde, the Herbalist, as follow:

> A lot of people plant without knowing the uses of what they planted. Sometimes they thought it's just [*ornamental plant*] flower and for beautification; when we see those plants, we go and ask them to give us.

He reinforced the difference between mystic use of herbs from non-mystic everyday use of plant materials for human wellbeing without spiritualizing nature, with another illustration that:

> If someone is having headache there is a gourd that is being used to remove the headache [*that*] works like a remote. Some charms work only as commandment, that is you don't have to use any plants.

Thus, herbal practices in Yoruba cities could be literally non-spiritual or otherwise. This places onus on the ecological designer to fully define the personhood and characterize the identity of the individual client during the design process at plot level. However, at the communal level, these users' characterization of communal ecological facilities would require rigorous procedure to achieve a balance between biophilic goals and spiritualisation of nature in ecological Yoruba urbanism. While these procedures are fully discussed in the next chapter of this book, it is important to highlight how such procedures could be integrated with the key debates in this chapter. These are the means of achieving balance between biophilic goals and spiritualisation of nature in ecological Yoruba urbanism. Firstly, both biophilia and spiritualisation of nature dominate the ecologies of Yoruba cities. Inclusive city planning in the region means that ecological ideologies would incorporate development ofguidelines for review of the existing building and planning regulations which have colonial origin. Secondly, human ecology is central to the urban history of Yoruba people. This makes them to be environmentally conscious and depends on their oneness with elements of nature in everyday life for wellbeing. Natural elements (soil, water, air, plants, and animals) define the utilitarian, aesthetic, and spiritual satisfactions of city dwellers in Yoruba region through access to ecosystem services. Thirdly, the ecological urbanism of Yoruba cities is crafted with mythological indigenous knowledge system. The ongoing urban renewal processes in the cities are driven by political will and could act as erasures to the eco-cultural foundations of the citiesif not situated within eco-cultural goals. The next chapter of the book focuses on the formulation of evidence-based design and planning strategies for translating the indigenous knowledge system of Yoruba people on urban ecosystem services into robust policy tools.

References

Adedeji JA (2020) Urban biocultural identity of Yorubas: intersection of philosophy and nature for wellbeing at Osun sacred grove UNESCO site, Osogbo, Nigeria. In: Cocks M, Shackleton C (eds) Urban nature: enriching belonging, wellbeing and bioculture, 1st edn. Routledge, London, pp 51–66

Adedeji JA, Fadamiro JA (2018a) Urbanisation forces on the landscapes and the changing value-systems of Osun sacred grove UNESCO site, Osogbo, Nigeria. *Landsc Res* 43(6):798–816

Adedeji JA, Fadamiro JA (2018b) Effect of contiguity of homes and gardens on residents' quality of life in Akure, Nigeria. In: Akinbamijo OB, Okoko EE, Omole FK, Popoola OO (eds) *The just city: poverty, deprivation and alleviation strategies*. Department of Urban and Regional Planning. The Federal University of Technology, Akure. Yemped Prints Ltd, Akure, pp 168–180

Adedeji JA, Fadamiro JA, Adedeji YMD (2016) Residential house gating and fence design: an overview of social inequality and urban insecurity in Ilorin, Nigeria. Cities 52:123–131

Adedeji JA, Fadamiro JA, Odeyale TO (2020) Design toolkits for campus open spaces from post-occupancy evaluations of federal universities in south-West Nigeria. Built Environ Proj Asset Manag 10(2):296–311

Anguelovski I, Roberts D (2011) Spatial justice and climate change: multiscale impacts and local development in Durban, South Africa. In: Carmin J, Agyeman J (eds) Environmental inequalities beyond borders: local perspectives on global injustices. MIT Press, Cambridge, pp 19–43

Araújo D, Brymer E, Brito H, Withagen R, Davids K (2019) The empowering variability of affordances of nature: why do exercisers feel better after performing the same exercise in natural environments than in indoor environments? Psychol Sport Exerc 42:138–145

Ayinla S, Ayodeji AA (2022) Re-interpreting Yoruba incantation: an example of Fryean apocalyptic and demonic imagery. Alter Native: An Int J Indig Peoples 18(1):75–83

Balogun OIH (2017) Orisa-Shakespeare: a study of Shakespeare adaptations inspired by the Yoruba tradition (Doctoral thesis, Victoria University of Wellington)

Bartholomew I, Bartholomew EP (2012) On earth as in heaven: ecological vision and initiatives of ecumenical patriarch Bartholomew, vol 3. Fordham University Press

Bauman WA, O'Brien KJ (2020) Environmental ethics and uncertainty: wrestling with wicked problems. Routledge, New York

Blackbourn D (2011) The conquest of nature: water, landscape, and the making of modern Germany. Random House, London
Briehl MM (2015) Oxygen in human health from life to death–an approach to teaching redox biology and signaling to graduate and medical students. Redox Biol 5:124–139
Cao X, Qiu J, Wang L, Zhou G (2022) An integrative model of tourists' pro-environmental behavior based on the dual path of rational planning and embodied emotion. Int J Environ Res Public Health 19(13):7910
Cocks M, Shackleton C (2020) Situating biocultural relations in city and townscapes: conclusion and recommendations. In: Urban nature. Routledge, pp 241–268
Conradie EM (2020) The four tasks of christian ecotheology: revisiting the current debate. Scriptura: J Biblical, Theol Contextual Hermeneutics 119(1):1–13
Cormenzana B, Fabregas F, Marinescu MC, Marrero M, Rueda S, Uceda-Sosa R (2014) An ontology for ecological urbanism. SUM+ Ecology. In: Workshops at the Twenty-Eighth AAAI Conference on Artificial Intelligence
Dada P (2017) I once died and woke up when about to be buried –Aroloye, Owa of Idanre, the Punch Newspaper of 29th April 2017.https://punchng.com/i-once-died-and-woke-up-when-about-to-be-buried-aroloye-owa-of-idanre/
Das M, Chatterjee B (2015) Ecotourism: A panacea or a predicament? Tour Manag Perspect 14:3–16
de Villiers R, Moolla FF (2018) *Natures of Africa: ecocriticism and animal studies in contemporary cultural forms*. Unisa Press, Pretoria
Decker H, Van Holde KE (2010) Oxygen and the evolution of life. Springer Science & Business Media, Berlin
Dobbs C, Vasquez A, Olave P, Olave M (2021) Cultural urban ecosystem services. In: Urban ecology in the global south. Springer, Cham, pp 245–264
el-Baghdadi O, Desha C (2017) Conceptualising a biophilic services model for urban areas. Urban For Urban Green 27:399–408
Engels F (1964) *Dialectics of nature*. Translated by Clemens Dutt. Progress Publishers, Moscow
Escobar A (1999) After nature: steps to an antiessentialist political ecology. Curr Anthropol 40(1):1–30
Esfandiar K, Pearce J, Dowling R, Goh E (2022) Pro-environmental behaviours in protected areas: a systematic literature review and future research directions. Tour Manag Perspect 41:100943
Farnham TJ, Kellert SR (2002) Building the bridge: connecting science, religion and spirituality with the natural world. In: Kellert SR, Farnham TJ, Farnham T (eds) The good in nature and humanity: connecting science, religion, and spirituality with the natural world. Island Press, pp 1–8
Gandy M (2015) From urban ecology to ecological urbanism: an ambiguous trajectory. Area 47(2):150–154
Garrett JK, White MP, Huang J, Ng S, Hui Z, Leung C et al (2019) Urban blue space and health and wellbeing in Hong Kong: results from a survey of older adults. Health Place 55:100–110
Giannico V, Spano G, Elia M, D'Este M, Sanesi G, Lafortezza R (2021) Green spaces, quality of life, and citizen perception in European cities. Environ Res 196:110922
Green MC, Haron M (eds) (2020) Law, religion and the environment in Africa, vol 7. African Sun Media
Hall C, Knuth M (2019) An update of the literature supporting the Well-being benefits of plants: a review of the emotional and mental health benefits of plants. J Environ Hortic 37(1):30–38
Hart J (2006) Sacramental commons: Christian ecological ethics. Rowman & Littlefield, Lanham
Heerwagen J (2009) Biophilia, health, and Well-being. In: Campbell L, Wiesen A, eds. Restorative commons: creating health and Well-being through urban landscapes. Gen. Tech Rep. NRS-P-39. US Department of Agriculture, Forest Service, northern Research Station: 38-57
Hunt CA, Harbor LC (2019) Pro-environmental tourism: lessons from adventure, wellness and eco-tourism (AWE) in Costa Rica. J Outdoor Recreat Tour 28
Kellert S, Calabrese E (2015) The practice of biophilic design. Terrapin Bright LLC, London

Kellert SR, Farnham TJ, Farnham T (eds) (2002) The good in nature and humanity: connecting science, religion, and spirituality with the natural world. Island Press, Washington DC
Kellert SR, Wilson EO (eds) (1993) The biophilia hypothesis. Island Press, Washington, DC
Keune H, Immovilli M, Keller R, Maynard S, McElwee P, Molnár Z et al (2022) Defining nature, Earth System Governance Series. University of Zurich, pp 25–42
Kingsley JY, Townsend M, Henderson-Wilson C (2013) Exploring aboriginal people's connection to country to strengthen human–nature theoretical perspectives. Adv Med Soc 15:45–64
Landscape Europe News letter (2015) Retrieved 19 August, 2015 from http://www.landscape-europe.net/images/LEN_January_2015.pdf
Langhorst J (2014) Re-presenting transgressive ecologies: post-industrial sites as contested terrains. Local Environ 19(10):1110–1133
Lin BB, Egerer MH, Ossola A (2018) Urban gardens as a space to engender biophilia: evidence and ways forward. Front Built Environ 4:79. https://doi.org/10.3389/fbuil.2018.00079
Lothian A (1999) Landscape and the philosophy of aesthetics: is landscape quality inherent in the landscape or in the eye of the beholder? Landsc Urban Plan 44(4):177–198
Löwy M (2017) Marx, Engels, and ecology. Capital Nat Social 28(2):10–21
Lumber R, Richardson M, Sheffield D (2017) Beyond knowing nature: contact, emotion, compassion, meaning, and beauty are pathways to nature connection. PLoS One 12(5):e0177186
Manzini NZ (2019) African environmental ethics as southern environmental ethics. In: Chemhuru M (ed) African environmental ethics. The international library of environmental, agricultural and food ethics, vol 29. Springer, Cham. https://doi.org/10.1007/978-3-030-18807-8_8
Matholeni NP, Boateng GK, Manyonganise M (eds) (2020) Mother earth, mother Africa & African indigenous religions. African Sun Media
Mowatt RA (2018) Understanding Ifá: inserting knowledge of an African cosmology in leisure studies and nature-based research. Leis Stud 37(5):515–532
Nasr SH (2006) Spirituality and science: convergence or divergence? The Essential Sophia, (Chapter 15), 12
Ogundipe A (2018) Esu Elegbara: chance, uncertainly in Yorùbá mythology. Kwara State University Press
Parsons G (2008) Aesthetics and nature. Bloomsbury Publishing, London
Preston JL, Baimel A (2021) Towards a psychology of religion and the environment. Curr Opin Psychol 40:145–149
Russell JA, Lanius UF (1984) Adaptation level and the affective appraisal of environments. J Environ Psychol 4(2):119–135
Ryan C (2019) Earth, air, fire, water and the big smoke. Urban paganism: ritual performance, identity construction and meaning making within and without the city. (Master's Thesis,. The Arctic University of Norway
Santos MM, Lanzinha JC, Ferreira AV (2021) Review on urbanism and climate change. Cities 114:103176
Sayem MA (2018) Religious understanding of nature and influence of geographical environment on shaping religious beliefs and practices within Christianity and Islam. KARSA: J Soc Islamic Culture 26(2):195–214
Scott C (1996) Science for the west, myth for the rest? The case of James Bay Cree knowledge construction. In: Nader L (ed) Naked science: anthropological inquiry into boundaries, power, and knowledge. Routledge, NY, London, pp 69–86
Scott M, Lennon M, Haase D, Kazmierczak A, Clabby G, Beatley T (2016) Nature-based solutions for the contemporary city/re-naturing the city/reflections on urban landscapes, ecosystems services and nature-based solutions in cities/multifunctional green infrastructure and climate change adaptation: brownfield greening as an adaptation strategy for vulnerable communities?/delivering green infrastructure through planning: insights from practice in Fingal, Ireland/planning for biophilic cities: from theory to practice. Plan Theory Pract 17(2):267–300

Shackleton C, Cocks M (2020) Urban nature and biocultural realities. In: Cocks M, Shackleton C (eds) Urban nature: enriching belonging, wellbeing and bioculture, 1st edn. Routledge, London, pp 1–25

Simpson AM (2022) Healing the planet: traditional spiritual beliefs and sustainable management of ecosystems in the Amazon Forest, Colombia. *Pract Theol* 1-13:432–444. https://doi.org/10.1080/1756073X.2022.2081290

Spirn AW (1997) The authority of nature: conflict and confusion in landscape architecture. Nature and ideology*:* natural garden design in the twentieth century, 249–261

Spirn AW (1998) The language of landscape. Yale University Press, New Haven

Spirn AW (2011) Ecological urbanism. In: Companion to Urban Design. Routledge, pp 614–624

Spring ÚO (2020) Earth at risk in the 21st century: rethinking peace, environment, gender, and human, water, health, food, energy security, and migration: with a foreword by Lourdes Arizpe Schlosser and a preface by Hans Günter Brauch, vol. 18. Springer Nature

Ssebunya M, Morgan SN, Okyere-Manu BD (2019) Environmental justice: towards an African perspective. In: Chemhuru M (ed) African environmental ethics. The international library of environmental, agricultural and food ethics, vol 29. Springer, Cham. https://doi.org/10.1007/978-3-030-18807-8_12

Studer HD (1998) Francis bacon on the political dangers of scientific progress. Can J Political Sci/Revue canadienne de science politique 31(2):219–234

Taylor JR, Lovell ST (2014) Urban home food gardens in the global north: research traditions and future directions. Agric Hum Values 31(2):285–305

Taylor JR, Lovell ST (2015) Urban homegardens in the global north: a mixed methods study of ethnic and migrant homegardens in Chicago, IL. *Renew Agric Food Syst* 30(1):22–32

Torabi M, Noori SM (2019) Religious leaders and the environmental crisis: using knowledge and social influence to counteract climate change. Ecum Rev 71(3):344–355

UNESCO (n.d.) Idanre Hills. Accessed from https://whc.unesco.org/en/tentativelists/5169/

Wilson EO (1984) Biophilia: the human bond with nature. Harvard University Press, Cambridge, MA

Wood RJ (2002) Scientific and spiritual perspectives of nature and humanity. In: Kellert SR, Farnham TJ, Farnham T (eds) The good in nature and humanity: connecting science, religion, and spirituality with the natural world. Island Press, pp 9–15

Zigale TT (2021) Climate change, pastoral livelihood vulnerability and adaptation strategies: a case study of Sitti zone, Somali regional state in eastern Ethiopia. (Doctoral dissertation,. University of South Africa

Zimmerer KS, Bassett TJ (2003) Approaching political ecology. Political ecology*:* An integrative approach to geography and environment-development studies, 1–25

Chapter 5
Conclusion: Design Strategies for Eco-cultural Cities

Abstract Yoruba cities are ecological urban settings of a large single homogenous ethno-linguistic group in Nigeria. These cities have eco-cultural qualities of ecological urbanism defined by Yoruba culture, local wisdom, and elements of nature rather than by built forms. As sustainable urban settings, they integrate human ecology with everyday cultural life to produce urban landscapes for living, working, and recreation spaces in traditional ways. Despite the volume of literature in this domain, there is dearth of understanding of how to translate the Indigenous Knowledge System (IKS) on ecological urbanism to formulate strategies that strive to balance the constantly evolving modernism with traditionalism in non-western contexts. The aim in this chapter is to define methods of achieving this form of modernized traditionalism despite modernity as the means of translating the IKS on the duos into design strategies for eco-cultural cities. The strategies are framework, model, and ethnographic design algorithms that are syntheses of the lived experiences of key informants that are presented and discussed in Chaps. 2, 3, and 4 of this book. At the macro level of the city, the chapter contains ecological planning model, framework, and design algorithm. They incorporate biotic, socio-cultural, and abiotic systems as ecological planning framework; varieties of green and blue spaces across city sectors as ecological planning model; and design algorithm for urban ecosystem services. The approaches to achieving eco-cultural homes at the micro level are also detailed in the chapter as generic and specifics of design algorithms for homegardens. These include design algorithms for curative medicinal, orchard, preventive medicinal, crops, and vegetables homegardens. These planning and design strategies are akin to ethnographic algorithms. The chapter contains reflections and recommendations that shed light on implications of the strategies for the formulation of building and planning policies.

Keywords Eco-cultural cities · Ecological planning · Evidence-based design strategies · Medicinal plants · Cultural ecosystem services · Human wellbeing

J. A. Adedeji, *Ecological Urbanism of Yoruba Cities in Nigeria*, Cities and Nature, https://doi.org/10.1007/978-3-031-34688-0_5

5.1 Planning and Designing Eco-cultural Cities

Chapters 1, 2, 3 and 4 of this book have detailed the manifestation of ecological urbanism in Yoruba cities with ethnographic lens that focused on urban ecosystem services of green infrastructure and Indigenous Knowledge Systems (IKS). The chapters are based on thorough considerations of the tenets of ecological urbanism through cutting-edge ontological, epistemological, and methodological groundings. However, a concluding quest presents a dialectic puzzle of balancing this traditionalism with modernism and the means of translating the IKS on the duos into evidence-based planning and design strategies for eco-cultural cities. The goal in this chapter is to synthesize the lived experiences of the participants meta-narrated in the core Chaps 2, 3, and 4 of this book to underpin how eco-city spaces could be designed to meet the needs for wellbeing of the Yoruba city dwellers. This fusion of ecology and culture to produce eco-cultural cities aligns with the vision of Arthur Tansley who wrote as far back as 1935 that:

> We cannot confine ourselves to the so-called "natural" entities and ignore the processes and expressions of vegetation now so abundantly provided by man. Such a course is not scientifically sound, because scientific analysis must penetrate beneath the forms of the "natural" entities, and it is not practically useful because ecology must be applied to conditions brought about by human activity. The "natural" entities and the anthropogenic derivatives alike must be analyzed in terms of the most appropriate concepts we can find (Tansley 1935, p. 304).

Therefore, eco-cultural cities are urban settings that have optimum balance of ecological tranquility and cultural requirements for the wellbeing of dwellers in practical senses. They combine nature and culture, not as binaries but as integrated and co-produced entities acting in urban spaces that produced them and they sustain simultaneously. As a double-edged approach to urbanism and city planning that leverages on complementarities between nature outside of humans and human nature, it fosters rewarding relationships between human and non-human ecosystems of cities. In view of its meshing of nature and culture together, eco-cultural strategy of city planning has to be place-specific to meet the target goal of human wellbeing. Even though modernism is a continuum, the establishment of contemporary eco-culturalism in line with this continuum would require the traditionalism of IKS to be interlocked with it at every temporal period to meet wellbeing of humans in the city. This is the essence of eco-cultural urban design and city planning. In this chapter, the subject is approached at the macro level of the city and the micro level of homes in the city spaces. The design considerations, factors, and indicators of eco-cultural cities include naturality, sustainability, place identity, place attachment, belongingness, tangible and intangible heritages, cultural value-systems, resilience, traditionalism, and aesthetics, among many others. They are used as crucibles for fathoming urban eco-cultural place-making. This approach is in agreement with Tandarić et al. (2020) who identify "place, people, past, practices, and purpose" as considerations for planning of urban cultural ecosystem services (p. 1).

Furthermore, ecological planning framework is a graphical or textual (or both) representation of the concepts and their links that make up the phenomenon of ecological urbanism and its thesis of eco-cultural cities for holistic understanding. It is essentially a practical framework that is conceptual and "defined as a network or a 'plane' of linked concepts that together provide a comprehensive understanding of a phenomenon or phenomena" (Jabareen 2009, p. 51). The constituting concepts of the framework "support one another, articulate their respective phenomena, and establish a framework-specific philosophy" (ibid, p. 51). The supporting philosophy for all the model, algorithms, and framework in this book is fully discussed in Chap. 1. The interwoven connection of the components of a framework is shown by their characteristic consistency, endoconsistency, and heterogeneity, despite being inseparable (Deleuze and Guattari 1994, p. 19). Baschak and Brown (1995) developed an ecological framework on the basis of scale for the planning, design, and management of urban river greenways. While planning considers a regional scale with focus on organism communities through ecosystem science, design is related to site-specific issues on vegetation type through naturalistic approach. In the case of management, the scale focus is on land unit at ecodistrict and ecoregion statuses through landscape ecology. These are meant to achieve a balance between biophysical and socioeconomic objectives. The ecological framework of Baschak and Brown (1995) incorporates assessment process, inventory, network structure, urban context, and greenways. It presents a robust approach to understanding urban ecosystem patterns to the lowest level of plant species and overall biodiversities. La Rosa et al. (2018) proposed a planning framework for urban green spaces that incorporate the demand of different social groups. The framework is divided into four processes. These include delineation of objectives, modelling of accessibility, analysis of results, and explanation of planning decisions. Through the perspectives of human ecology and ecology of green spaces, it accounts for social, spatial, and locational indicators of ecosystem services.

The foregoing suggests the possibility of translating the lived experiences on local wisdom and IKS of urban ecosystem services of the participants in the study presented in this book to evidence-based design strategies for achieving eco-cultural city planning. Since city planning and urban design require contextual consideration and place specificity, the approaches of the Global North are not agreeable with the cultural milieu of Yoruba people. For this reason, cultural congeniality accounts for the formulation of the eco-cultural planning and design strategies described in the Figures that follow. They are situated within the epistemic limits of the key results in this study and therefore non-exhaustive of the Yoruba cities. However, the diverse socioeconomic backgrounds of the participants and the expansiveness of their geographical spread within the Yoruba region assure of a robust level of the possibility of generalization to an acceptable status. The underpinning justifications for this generalization are detailed in Chap. 1 of this book and are not intended to be repeated here. However, it is necessary to recall that these justifications are based on the homogenous nature of the morphology of Yoruba cities. This homogeneity was precipitated by uniformity of some factors of development that are common to the cities. Summarily, these are common culture and local wisdom of Yoruba people that are reflected in the spatial arrangement of the cities, and common urban planning and development history because they have been governed together as a political entity and under common urban planning legislation. Therefore, there are nine strategies altogether and they are grouped into two, city and home bases. The city basis is considered in this section and Sect. 5.2 while the home basis is considered in Sect. 5.3.

Figure 5.1 shows an ecological planning framework for Yoruba cities. The ultimate goal is ecological urbanism evidenced in eco-cultural cities. The framework contains three systems. These are biotic, socio-cultural, and abiotic systems. While these are generic, their components coalesce in unique cultural ways in Yoruba urbanism based on biophilia and sacredness. The biotic system consists of the flora and fauna components. The groundcover, shrubs, and trees are primarily material in non-sacred spaces but are believed to be part of the cosmos in sacred spaces where they are tendered and conserved through mythological approaches like taboos and folklores. They constitute the urban green spaces of street trees and in traditional public squares, public parks and gardens, homegardens, and urban groves. They offer varieties of urban ecosystem services toward the realization of critical sub-goals of the 2030 agenda of the United Nations Sustainable Development Goals. The fauna component consists of human and non-human multispecies operating at two levels, physical and metaphysical. The human category is essentially a complex web of socio-cultural system operating with the interplay of kinship, traditional belief-system and religion, language (including divination and incantations), economy, and all other socioeconomic factors.

Therefore, eco-cultural cities are all-inclusive of the social profiles of the dwellers. Accordingly, respondents in this study are from varying social contexts. The social contexts include gender, age, education, health, nutrition, security, and housing. In terms of gender, the representation of males and females are profuse. The ruling class includes traditional, religious, and political rulers. The abiotic system is a complex web of lithosphere, hydrosphere, and atmosphere with their cosmological intricacies. In particular, some rocks and water bodies like oceans, lagoons, lakes, and rivers in the cities are believed to be abode of spirits, and offer many other cultural, provisioning, regulating, and supporting ecosystem services.

The ecosystem services in the framework could be achieved through an algorithmic design process as developed in Fig. 5.2. The ecosystem services' design algorithm is a complete loop with a start-and-end procedure. It incorporates a number of system steps. The first step is to select the desirable ecological space. This could be an urban grove, homegarden, public park or garden, institutional park or garden, lagoon or lagoonfront, river or stream. While urban groves, lagoons, rivers, and streams are naturally in existence and cannot be designed but identified and characterized for functional inclusiveness, others could be designed. The next step is to identify the intended ecosystem services' goals of the ecological space with supporting as the first, followed by cultural, provisioning, and regulating ecosystem services in order. Among the cultural ecosystem services' goal, urban

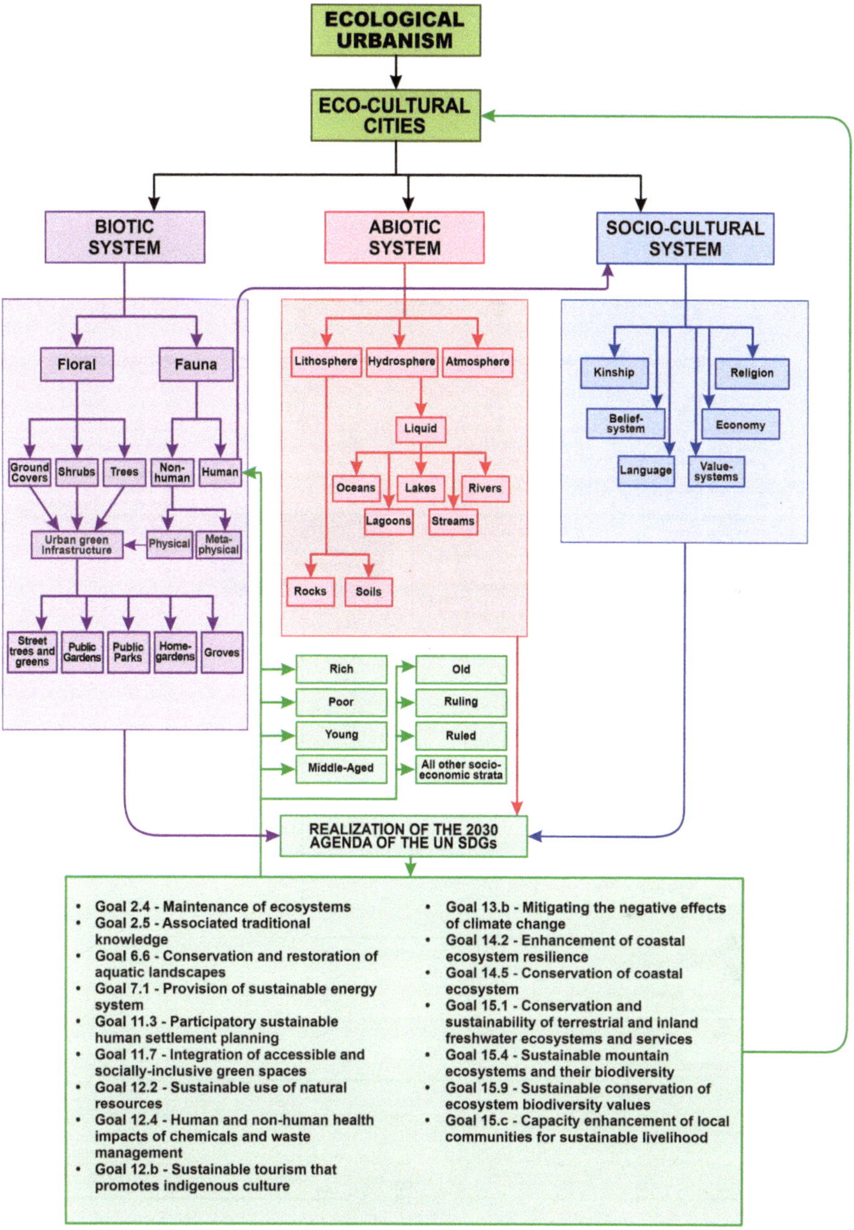

Fig. 5.1 Proposed ecological planning framework for Yoruba cities

groves offer spaces for traditional religious worship, cult systems and divination, understanding Eco-theology, and spiritual bathing in their water bodies. Urban groves also offer ecosystem services like urban ecotourism that results in

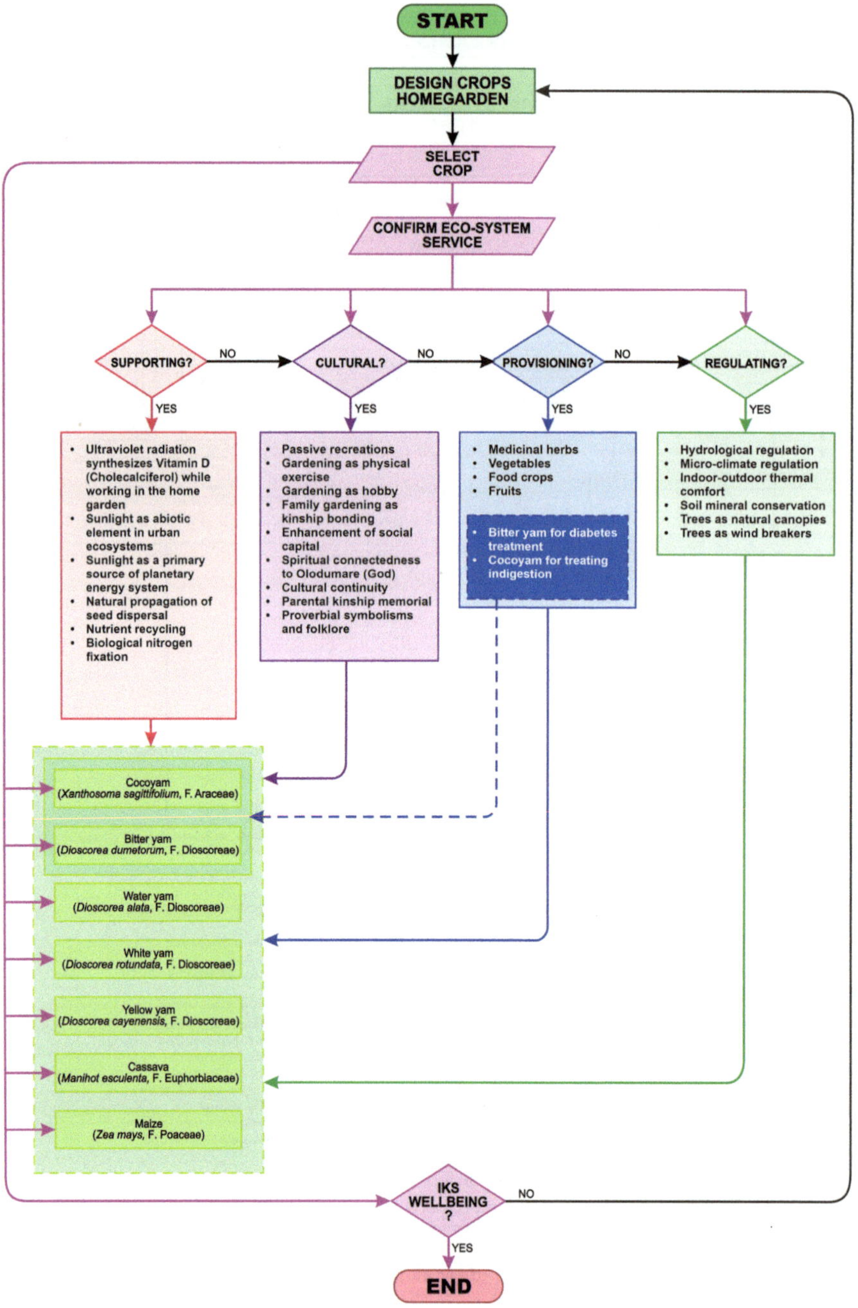

Fig. 5.2 Proposed ecosystem services' design algorithm for Yoruba cities

reconnection with nature and respect for local culture. All other spaces offer recreation, nature appreciation, adventure, sightseeing, education and research. Others include acting as vital component of social capital, image making and city branding, ecological memorials, providing employment opportunities, natural canopies, and boosting the hospitality industry. Many provisioning ecosystem services are also germane for identification, characterization, and functional inclusion. These are provisioning of medicinal herbs and concoction, spiritual bathing, building and furniture materials, industrial base materials including microbiomes, building sand, and fishing. Furthermore, desirable regulating ecosystem services include green spaces acting as carbon sink, reduction of global warming, reduction of air pollution, microclimate regulation, soil cohesion, soil mineralization and fertility, and recycling of soil nutrients. Supporting regulatory services to be pursued include protection of wildlife species and biodiversity conservation, sustenance of ecosystem food web, human metabolism and therapeutic mental health, and natural plant pollination.

Within the contexts of Figs. 5.1 and 5.2, eco-cultural urban structure combines ecology and culture to drive urban resilience with the goal of achieving sustainability through the local communities. The frameworks are achievable, driven, and reinforced through a participatory design modelling of bottom-up approaches that take cognizance of fine-grained cultural lifestyles of traditional city dwellers through their IKS. The structure of Yoruba cities is made up of two categories of components, that is, biological and physical. The biological components include flora, fauna, and all forms of life. The physical components include soil, water, air, climate, and topography. Eco-cultural urban structure therefore requires water networks, green infrastructure networks, transportation and commuting networks, urban metabolisms (combining urban ecology, political ecology, and industrial ecology). Specifically, urban metabolisms have been discovered to be supported by the ecosystem services of green infrastructure (Perrotti and Stremke 2020). The network of ecosystem services is an urban structure that enhances great reduction in urban carbon footprint and energy saving in Yoruba cities as included in Fig. 5.2. It also supports sustainable urban metabolisms through responsive harvesting and use of natural resources where there is optimum balance between "inflows and outflows" (p. 680). The carbon-sequestration role of urban green infrastructure (UGI) makes them ecological while still meeting the cultural ecosystem needs of the city dwellers in multiple ways through IKS. This necessitates the implementation of UGI agenda in the planning and design of eco-cultural cities. The presence of UGI network in eco-cultural settings has been linked to the "flows and stocks of energy, materials and substances" as important components of urban energy metabolism systems (p. 681).

With these considerations, the design or redesign of eco-cultural cities to take cognizance of IKS would require the implementation of the following components of urban ecological systems (Grimm et al. 2000): socioeconomic drivers including flow of information, cultural values and perception, institutions and organisations; patterns of human activities including demographic patterns, economic systems, power hierarchies, land use and management, designed environment; and ecosystem dynamics. Others are biogeophysical drivers including flow of energy, re/cycling of matter, flow of information; and patterns and processes of ecosystems including primary production, populations, organic matter, nutrients, and disturbances (p. 574). These design considerations are much similar to what Qtaishat et al. (2020) summarized as eco-cultural sustainability with four major components as cultural, environmental, social, and economic aspects. They also align with the United Nation's SDGs and Agenda 2030.

5.2 Evidence-Based Design Models for Ecological Cities

Design models are applicable to the delivery of ecological cities for human wellbeing. These are represented or conveyed as ecological planning model. Models are ways of elucidating how something works by conveying

> Epistemic commitment to a comprehensive, systematically linked body of information that is grounded in fact, is duly responsive to reasons or evidence, and enables nontrivial interference, argument, and perhaps action regarding the topic the information pertains to (Elgin 2017, p. 44).

In particular, ecological planning model is a tool that allows a comprehensive overview of the components, processes, actions, and actors for the delivery of sustainable ecological regions at macro scales and cities at micro levels. It has its origin with the work of Ian McHarg (1969) *Design with nature* and has been popularized as a path of wisdom for realizing sustainability with the dictum, "following nature's lead in planning and design" (Yang and Li 2016, p. 22). The model frames city landscapes as socio-ecological settings of humans and ecosystem (Hersperger et al. 2021), and therefore requires deep insight into these processes. The idea of Ian McHarg (1969) has found expression in urban design by integrating it as urban ecology (Douglas 2019) in his real life projects. Johnson et al. (1979) proposed an ecological planning process culminating into land use plan production. The process consists of inventorizing and analyzing landscape constituents, interpreting the constituents to identify opportunities and constraints, determining and mapping of intrinsic appropriateness for intended land uses. Others are, confirming areas that are suitable for more land uses, analyzing the needs and longings, inventorizing and analyzing other intervening factors like socioeconomic and political variables. With a more broken-down procedure, Steiner (2008) proposes a regional ecological planning model that has 11 action components. These include identification of problem or opportunities, establishment of goal, inventory and analysis at regional level, detailed studies, planning concepts, landscape planning, involvement of citizens and education, detailed designs, implementation of plan and design, and administration (p. 11). These models are generic and their processes could fit into many contexts. However, in the case of Yoruba cities, there is a principle of environmental cultural determinism (Rapoport 1969; Adedeji and Amole 2010) that dictates the urban morphologies as shown in Fig. 5.3.

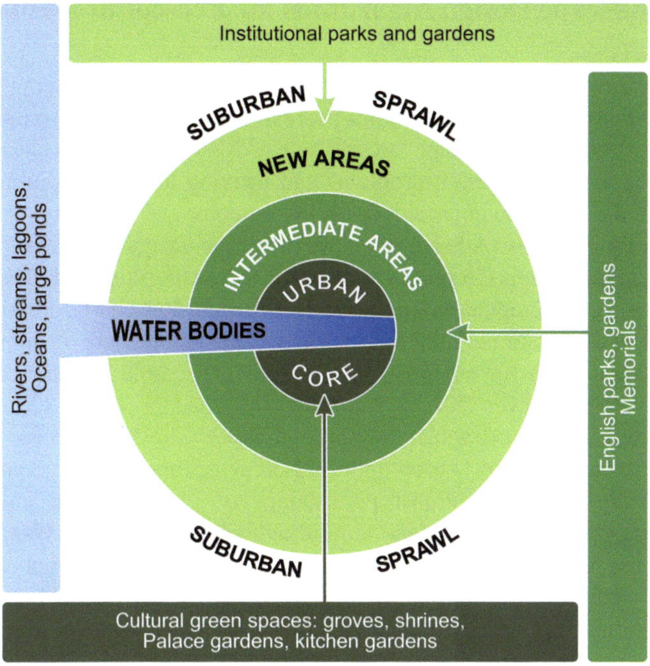

Fig. 5.3 Proposed ecological planning model for Yoruba cities

Figure 5.3 is a proposed generic ecological planning model for Yoruba cities. It shows four concentric city sectors. These are urban core, intermediate areas, new areas, and suburban sprawl. Depending on different hydrological character of each city, the four sectors are traversed by different typologies of water bodies. These include rivers, streams, lagoons, ocean, and large ponds. The urban cores are generally home to cultural green spaces including groves, shrines, palace gardens, and utilitarian homegardens like kitchen gardens. The intermediate areas are home to English-types of parks, gardens, and memorial green spaces, all of which are non-traditional. The new areas of the cities contain educational institutions of higher learning, specifically Universities, among others. This model frames the eco-city structure of Yoruba cities. The "landscape model" of Osogbo was developed by Adedeji and Fadamiro (2018, p. 813) through monarchical historiography. The model has four concentric but non-symmetrical layers of landscapes. These are natural, traditional, modern, and post-modern landscapes. Compared with Fig. 5.2, the model is interpretable as follows: the natural landscape is at the core areas, the traditional landscape is at the intermediate areas, the modern landscape is at the new areas, and the post-modern landscape is at the suburban areas. Though produced through the concept of cultural determinism (Rapoport 1969), the Osogbo landscape model compares with Burgess's (1925) concentric model. Similarly, the model in Fig. 5.3 compares with that of Burgess's (1925) but only in morphology.

5.3 Landscape Design Algorithms for Eco-cultural Homes

Design algorithms are logical qualitative or empirical problem-solving procedures through stepwise dialogues with input-process-output continuum to optimize best design decisions. In landscape architecture design, they offer clearly defined procedures for solving design problems and arriving at varying optimum design decisions and accompanying multiple quality outcomes successfully. They are akin to set of design toolkits (Adedeji et al. 2020) for the design of outdoor spaces and ethnographic algorithms (Gobo 2018; Seaver 2017). Algorithms could be expressed in plain languages, computer programming languages, flow charts, drakon charts, or pseudocodes. Algorithmic thinking of landscape architecture design has been judged to be suitable for exploring the patterns observable in nature and culture (Claghorn 2018; Seaver 2017). Shafizadeh-Moghadam et al. (2021) explore selection algorithm to extrapolate urban landscapes with the outcome of a future prognosis. Li and Xu (2021) used optimisation algorithm to study plant landscape planning and discovered utmost percentages of planning path and aesthetic performance. "Artificial bee colony algorithm" was developed by Wang and Zeng (2020) for the planning of green spaces of urban gardens having "high optimisation efficiency" (p.393). In a landscape study by Haase et al. (2019) "house-attached front and back yard green derivation algorithm" was used to examine home vegetation and discovered that there were more home greens "than public green space" in Leipzig city (p. 44). It is for these reasons that qualitative algorithms are suitable for suggesting landscape design praxes of eco-cultural homes.

As part of the syntheses of the results of this study, six qualitative ethnographic algorithmic thinking about landscape design and specifications for eco-cultural homes in Yoruba region are developed. These include design algorithms for general homegarden, curative medicinal homegarden, orchard homegarden, preventive medicinal homegarden, crops homegarden, and vegetables homegarden. Figure 5.4 shows the Homegarden design algorithm with start-process-end procedures. While the entire procedures lead to eco-cultural home, there is need to set desirable goals in pursuant of requisite ecosystem services which determine the type(s) of homegarden to design. While majority of the homegarden types could produce multiple ecosystem services, each type is best for particular ecosystem service. In this scale, orchard homegarden takes the lead by being most efficient for all the four ecosystem services – provisioning, regulating, supporting, and cultural. Orchard homegarden could also double as medicinal homegarden depending on the types of fruits specified. While vegetable homegarden largely supports provisioning ecosystem service, it could also double as medicinal homegarden depending on specifics and range of vegetables specified. The ecosystem service goal of crops homegarden is primarily provisioning but has additional advantage of regulating and less of others except for the cultural basis of gastronomic lifestyles. It is envisaged that properly designed homegarden should naturally achieve the goals of human wellbeing in the algorithmic process.

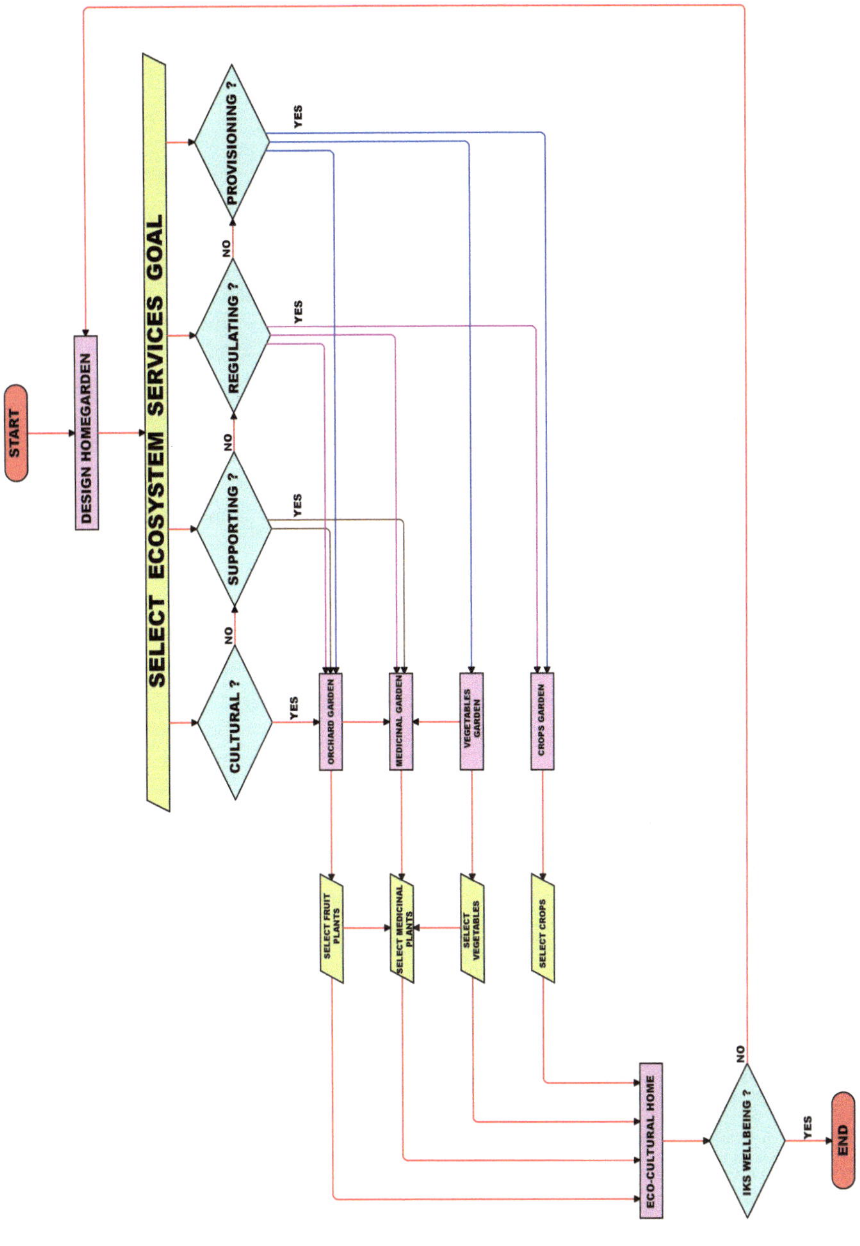

Fig. 5.4 Proposed homegarden design algorithm for Yoruba cities

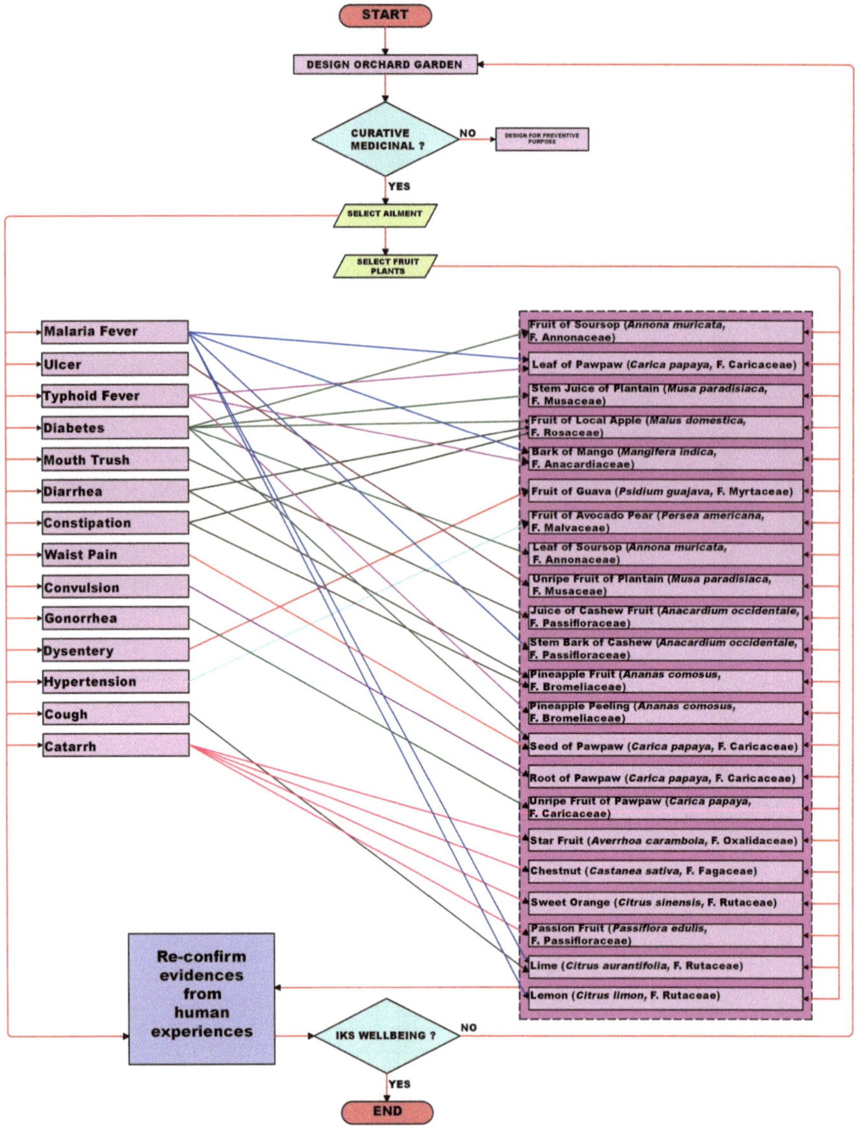

Fig. 5.5 Proposed curative medicinal homegarden design algorithm for Yoruba cities

Figure 5.5 shows the curative medicinal homegarden design algorithm. The selection procedure ensures adequate matching of specific plants with specific ailments. While most plants have multiple ethnomedical ecosystem services, some exhibit their curative properties against a particular ailment higher than others. In addition, some medicinal plants are suitable for everyday living because they act as impetus to human wellbeing without necessarily being in pursuant of curative goals. Of all the plants that could be specified in the landscape design, life plant (*Bryophyllum pinnatum*, F. Crassulaceae) is obviously indispensable for being suitable for numerous ailments. Based on the evidences gathered in this study, medicinal homegardens that are designed with plant specifications in the algorithm should essentially lead to the fulfillment of IKS for human wellbeing aside from intervening factors that could be medical or non-medical that are outside the scope of the study.

Figure 5.6 shows the algorithmic design thought for orchard homegarden. It details the plant selection procedure for ethnobiological curative purpose aside from the orchard being a vibrant source of assorted fruits for preventing ailments in human body and enhancing wellbeing. In addition to guiding the specification of relevant fruit plants, the algorithm also shows that all parts of most fruit plants have traditional medicinal values. The wide range of fruit trees and the canopy foliage system of some of them also assure other numerous ecosystem services in the home environments.

Figure 5.7 shows the preventive medicinal homegarden design algorithm with a start-process-end procedure. The selection process for the garden plants involves a mix of fruits, vegetables, crops, and utilitarian medicinal plants with majority of the plants providing anti-oxidant activities in human body. Their arrays of preventive health benefits are all-embracing as leaves, roots, barks, fruits, nuts, and stems of the plants are ethnobotanical items for provisioning ecosystem services. Specifying these plants in the landscape plan of homegardens not only ensures human wellbeing but also supplies other numerous urban ecosystem services that enhance indoor and outdoor environmental qualities.

Figure 5.8 shows the crops homegarden design algorithm. With start-process-end procedure, it allows specification of a number of food crops discovered to be recurring in the cities according to the results of the study. Though largely for provisional ecosystem services of staple foods, the food crops offer other ecosystem services relatively. The provisioning ecosystem services include ethnomedicinal herbs, vegetables, food crops, and fruit crops. Since bitter yam was identified during the study to be effective for the traditional treatment of diabetes and cocoyam for curative and preventive treatment of indigestion, they are important for inclusion in homegarden planning. The supporting ecosystem services to pursue include synthesis of Cholecalciferol. This is through early morning sun as natural 'gift' from nature for tending it through gardening, energy source to the floral system through the solar energy, natural seed dispersal as a means of propagation, recycling of soil nutrients for the floral lives, and nitrogen fixation through biological mechanisms. Cultural ecosystem services to pursue include the recreational ecosystem services of the ecological space. Others are the goal of the gardening whether as a hobby or not and who are involved in the gardening process. If family members are involved, this could enhance kinship bonding or contribute to social capital of the community if non-family members are involved. It is also important to indicate if individuals might be emotionally attached to the spaces through spiritual process or how the spaces stimulate continuity of cultural lifestyles, or enhance reconnection to childhood memorial. It is necessary to highlight the extent to which proposed ecological spaces could function figuratively as a symbol for indigenous wisdom, folklore and urban myths. All desirable regulating ecosystem services on hydrology, micro climate, thermal comfort in the indoors and outdoors

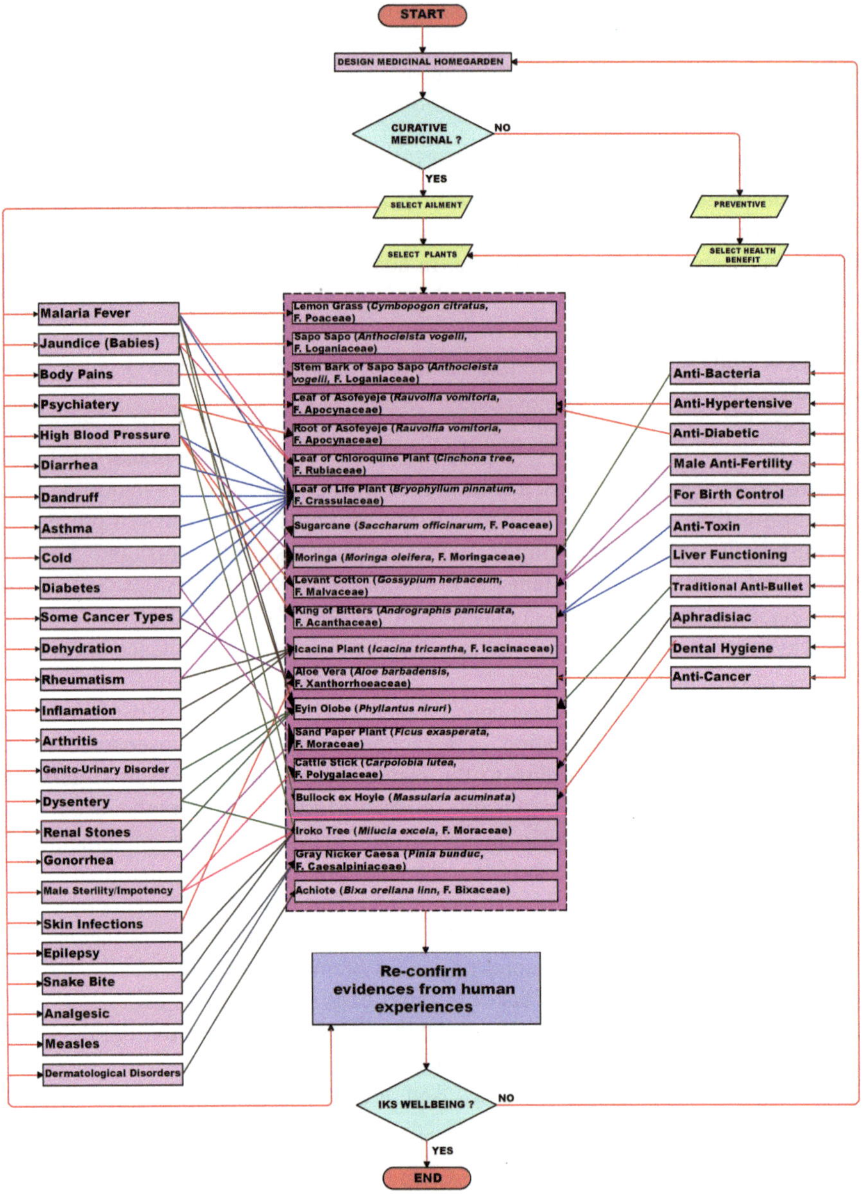

Fig. 5.6 Proposed orchard homegarden design algorithm for Yoruba cities

of built forms, soil mineral conservation, and protection of built-forms from wind damage through wind-breaking trees should be focused as design goals.

Figure 5.9 shows the vegetables homegarden design algorithm. With start-process-end procedure, the algorithm is suitable for the best choices of gastronomic

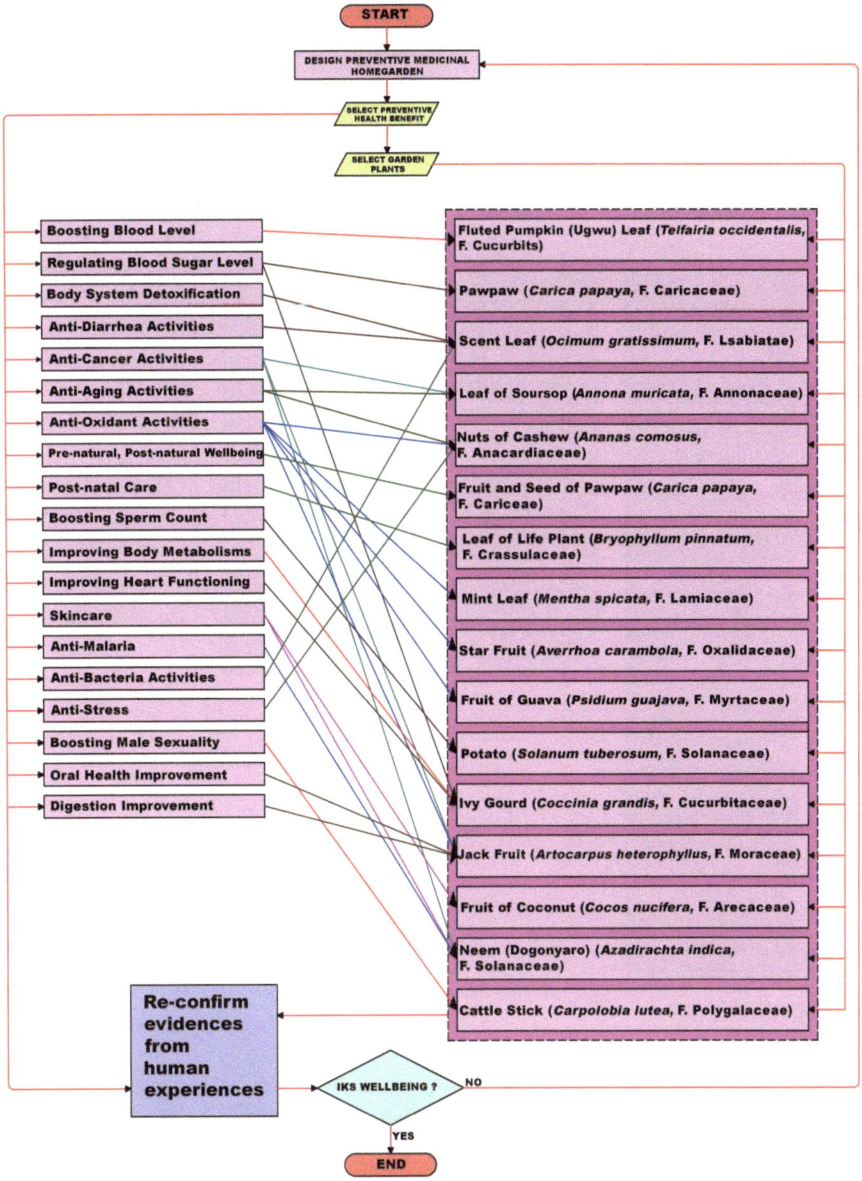

Fig. 5.7 Proposed preventive medicinal homegarden design algorithm for Yoruba cities

vegetable plants that double as curative medicinal herbs. Jerusalem artichoke (*Helianthus tuberosus L.*, F. Asteraceae) appears to be suitable for treating more ailments than others and could be used to treat bone fracture, skin wounds, swelling, diabetes, and rheumatism. Conversely, anaemia appears to have more varieties of medicinal vegetables than others for its cure. It could be treated with fluted pumpkin

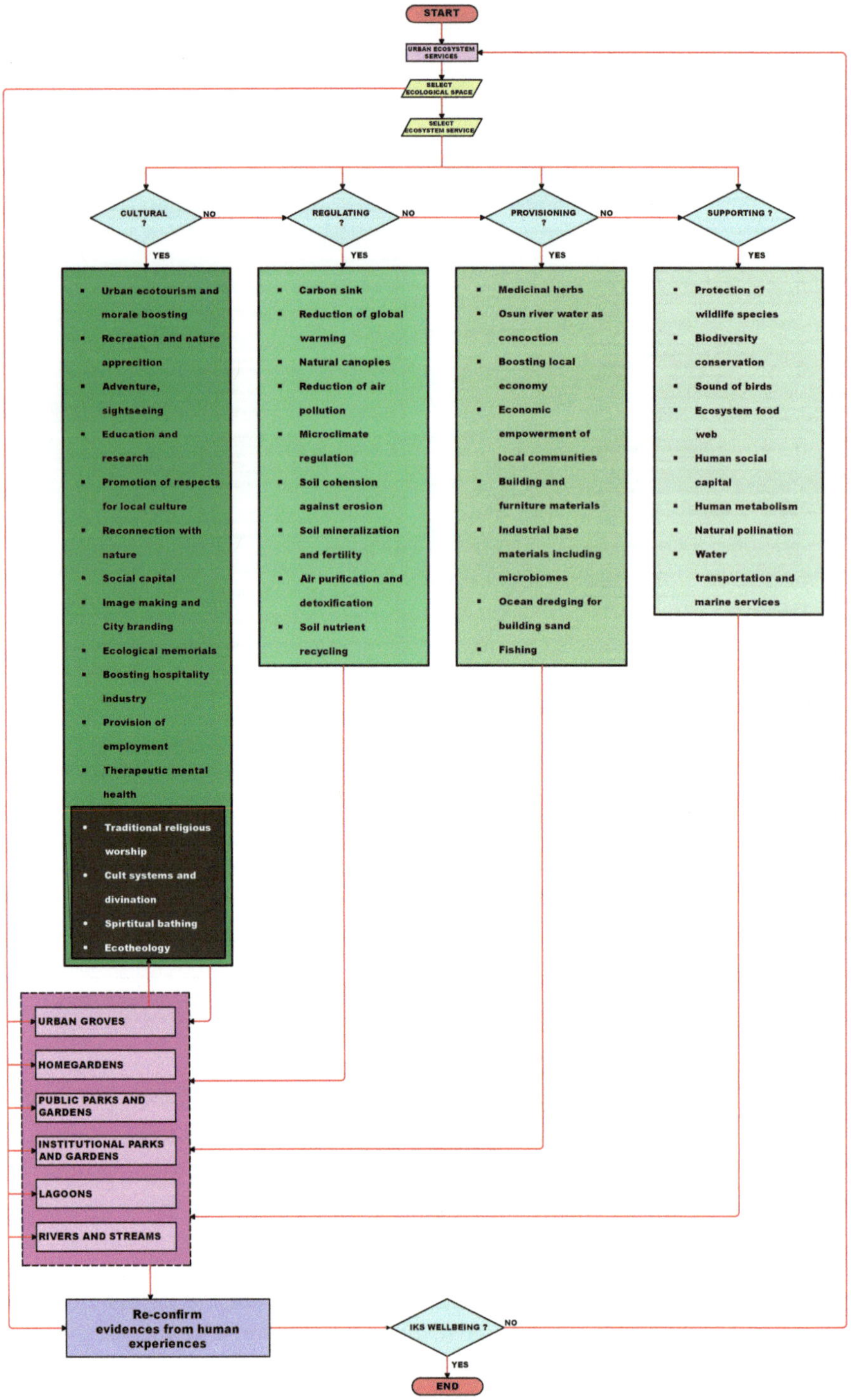

Fig. 5.8 Proposed homegarden design algorithm for Yoruba cities

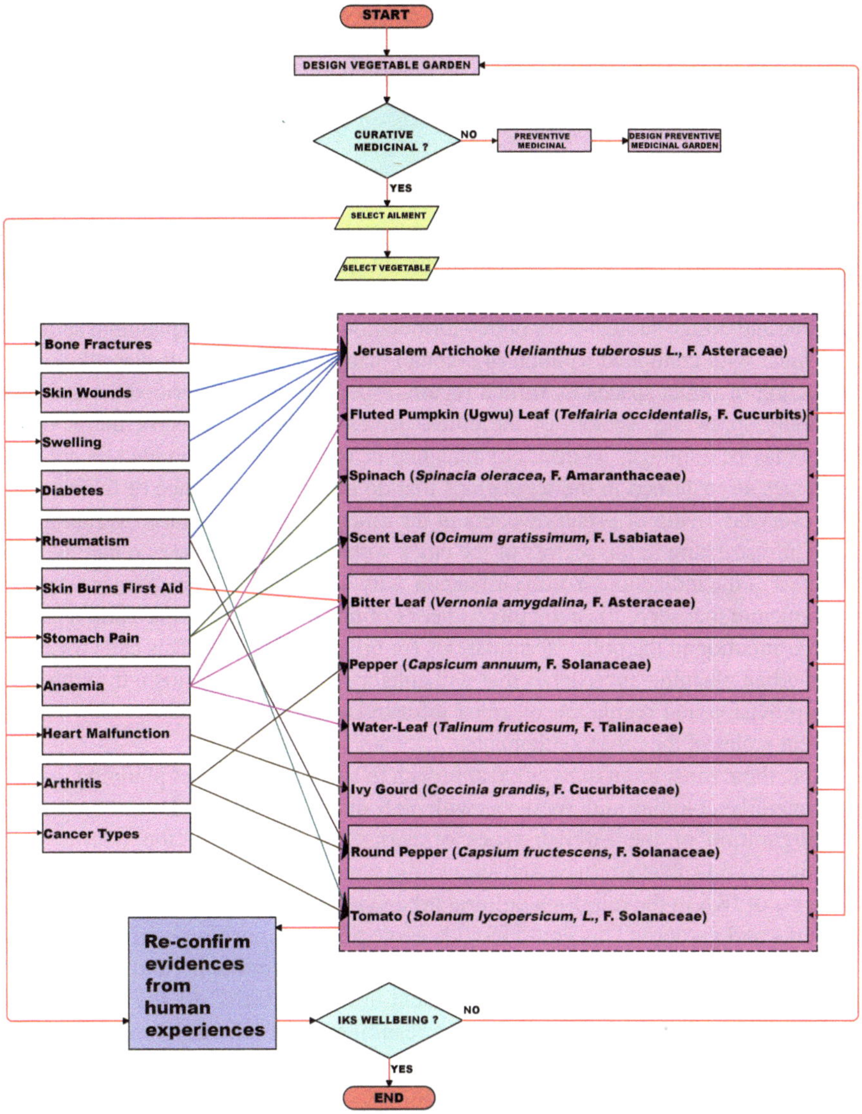

Fig. 5.9 Proposed vegetables homegarden design algorithm for Yoruba cities

(*Telfairia occidentalis*, F. Cucurbits), spinach (*Spinacia oleracea*, F. Amaranthaceae), bitter leaf plant (*Vernonia amygdalina*, F. Asteraceae), and water leaf (*Talinum fruticosum*, F. Talinaceae). Correct landscape plant specification, implementation, and consumption of plants from vegetable gardens in urban home environments should normally lead to realization of human wellbeing through this IKS.

On the whole, Figs. 5.1, 5.2, 5.3, 5.4, 5.5, 5.6, 5.7, 5.8, and 5.9 are a complete set of design strategies and toolkits that are not mutually exclusive of each other for achieving IKS of human wellbeing in eco-cultural cities.

5.4 Reflections and Recommendations

The negative effects of modernisation on the eco-cultural milieu of Yoruba cities and homes could be ameliorated by applying the formulated evidence-based ecological planning and design strategies for policy formulation on building and planning approval processes. With continued triumph of colonial planning ideologies driving the formal production of urban spaces in Yoruba region, especially in the intermediate and new (generally sprawling) sectors of the cities, achieving city-round eco-cultural urban milieu may be a mirage. To-date, city planning in Yoruba region is guided by colonial ideologies as enshrined in the regulations that do not take cognizance of the eco-cultural everyday living of urban dwellers in the cities. This has produced variations of green infrastructure in the sectors of the cities with only the core areas retaining their identities of the indigenous Yoruba urbanism. City-scale public open spaces and green infrastructure that serve the eco-cultural needs of urban dwellers in the Yoruba region deserve attention in the policy formulations for urban governance. This calls for innovative urban planning approaches that will enhance ecological urbanism through adequate provision and regulation of green infrastructure in Yoruba cities, based on the different scales of the urban settlements.

First, there must be an innovative approach to balance modernist planning, which are generally colonial and Western, with IKS at the city-scale. This requires the reconfiguration of the English types of parks and gardens in the intermediate (transition) and suburban (new and sprawling) zones of the cities by introducing the elements of IKS in the religious-cultural urban green spaces into the morphology of the parks and gardens. This could include ecological composition landscape design and morphological elements like form, texture, patterns, structures, and biomimicry, among others, that bear the intangible identities and 'spirits' of the eco-cultural green spaces in the urban core areas. The resulting green infrastructure spaces that would be developed will be for non-religious ecosystem services and they would not exhibit characteristics of the urban groves. Within this context, the English-types of green infrastructure would continue to function as recreational spaces that are eco-culturally branded for numerous ecosystem services.

Second, operational land use planning and regulations in Yoruba land lack principles of ecological landscape planning and design. Thus, specifications on the required kind of landscaping for the open spaces around buildings are lacking. For this reason, individuals apply their IKS to produce homegardens that are informal. This makes the sprawling suburban residential neighbourhoods lack coherence in terms of image and aesthetic appeal that should normally result from formal planning. Many home developers also indiscriminately pave the whole air-spaces around their buildings thereby contributing to thermal challenges in the neighbourhoods.

There are so many diversities of homegardens' configurations that are at variance with logics of aesthetics and principles of ecological landscape design due to lack of building and planning regulations on how best to produce eco-cultural green infrastructure. This is also an outcome of the lack of principles of ecological landscape planning and design like repetition, balance, proportion, order, unity, and complexity (Yılmaz et al. 2018) in the cities. These challenges are prevalent in the sprawling suburban areas and new sectors of the cities and shows loss of the IKS on green infrastructure unlike in the core areas where eco-cultural patterns are still surviving. If the principles of ecological landscape planning and design are situated within the IKS for producing eco-cultural spaces in the cities, the outcome would be outstanding.

Urban planning response in this regard is to revise planning regulations across the states of Yoruba region with particular specifications on landscaping parameters for vegetal and floral materials to guide home developers on how the mandatory air-spaces could be planted for human wellbeing in the urban ecosystems. This would necessitate relinking with the ethos and value-systems that are culture-based solution to the ecological challenges and composed of complete connection and interaction of human and nonhuman nature. As such, the eco-cultural principles in landscape production will be entrenched in urban residents and home developers, especially in the new and suburban zone of the cities. This will make the undeveloped areas or plots to be designed and developed with ecological consideration rather than being ignored as as left-over spaces either untreated, generally paved with concrete, or bare. The attempt will avoid ecological dislocations, development of thermal hotspots or heat islands, and loss of eco-cultural concern. Therefore, the design of home landscapes in eco-cultural cities would benefit maximally from the knowledge of the cultural lifestyles of the urban dwellers and how they could bring their IKS to bear on their modern home environments. This would also satisfy the requirements of modernist planning and design through the IKS in new and suburban areas of the cities. The ongoing politically motivated urban renewal of the core areas of the cities would also benefit from such process for all-inclusive city re-planning that supports conservation of tangible and intangible heritages.

It is necessary to recall that the traditional housing forms in the cities are dominated by the courtyard multi-dwelling buildings that automatically include green courtyards for multiple ecosystem services. These are huge patrilineal communal settings of extended families with many units of related nuclear families. They still exist in the core areas of the cities studied. Their courtyards satisfy basic eco-cultural needs of dwellers by incorporating gardens and green spaces for immediate access to medicinal and culinary plants. The courtyard gardens could be in the form of kitchen garden, or open recreation space. Despite the single-family dwellings that resulted from modernisation and westernisation, eco-cultural green spaces could be incorporated into the building and planning approval framework in the Yoruba cities of today. In view of the extinction of the multi-family courtyard system due to a new and current sociocultural lifestyles of individualism, air-spaces around the buildings could serve, especially backyard gardens with a similar eco-cultural IKS and values in the mode of modernism.

Since homegardens are central to the livability of the home environments in Yoruba cities, the advocated changes to the policy framework on building and planning approval processes are part of the life-changing strategies as Henry Lefebvre (Lefebvre and Nicholson-Smith 1991) argues that we must first change space to change life. He emphasized the notion that space is not given but produced to achieve place which is an ecological act "based on the relations of organisms to one another and to their surroundings" (Prokopenko 2020, p. 257). Eco-cultural homegardens are achievable by specifying requisite trees, shrubs, climbers, grass, and other vegetal materials that supply many ecosystem services in the landscape plans. As shown in the proposed ethnographic planning and building ecological landscape design algorithms, model and framework specification of plants that have medicinal values for different ailing health conditions and medicinal plants for preventive treatments are crucial, based on cultural belief systems. Because of the tropical climate of the Yoruba region, annual, biennial, and perennial medicinal plants could be planned for city-scale implementation and specified for the homegardens. In multi-floor dwelling apartments, hydroponic strategies could be explored by using potted plants in home verandahs, balconies, and lobbies. This is highly admissible for everyday medicinal plants like Alovera and indoor ethnobotanical plants. Then, it means required architectural spaces to contain these 'hanging' homegardens should have been provided during the necessary stages of formal architectural designs. Also, since ornamental gardens are not favourites of the indigenous Yoruba people, they should normally be excluded in the planning of their eco-cultural homes. Architects and urban designers are not expected to impose modern design ideologies that are anti-cultural on Yoruba clients but rather produce designs to satisfy eco-culturalism of Yoruba people. Eco-cultural homegarden landscape planning framework would benefit from a planting layout of medicinal, vegetable, orchard, and urban agriculture sections. In some instances, a merger of medicinal-vegetable section would be necessary where medicinal vegetables are largely desirable. These classifications are not exclusionary since most of the eco-cultural homegarden plants are medicinal.

The recurrence of medicinal plants in the study is evidence that IKS of human wellbeing is crucial to eco-cultural homegardening in the Yoruba region. The ecosystem services of these plants are not only most desirable but they are the reasons why majority engage in the planting of homegardens. In effect, there would be no homegarden without medicinal plants or gastronomic plants that serve medicinal purposes. For this reason, they need to be specified in the landscape plans of homes and as herbal shrubs in the region. Prominent among these is bitter leaves which is both herbal and of gastronomic ecosystem services. Another dimension of ecosystem services that emerged from the study is the use of medicinal plants for home landscaping and not necessarily in a garden. The flowering qualities of medicinal plants make them suitable for this ecosystem service. For instance, Jerusalem artichoke is suitable for landscaping being a source of numerous phytochemicals and has advantage of survival during the dry season and other harsh environmental conditions. A typical homegarden landscape plan should contain soursop considering its multifarious health benefits. Lemon grass and chloroquine plant could be planted as hedges, while Miracle plant is almost non-negotiable for inclusion in plant specification schedules for homegardening and home landscaping. Others are Mint leaf (*Mentha spicata*, F. Lamiaceae), Star fruit (*Averrhoa carambola*, F. Oxalidaceae) and Chestnut (*Castanea sativa*, F. Fagaceae), among many others, as included in the ethnographic design algorithms. King of bitters plant (*Andrographis paniculata*, F. Acanthaceae), and Aloe vera (*Aloe barbadensis,* family Xanthorrhoeaceae) are inevitable.

In view of the evidences, Ivy gourd (*Coccinia grandis*, F. Cucurbitaceae), Tea leaf (lemon grass) (*Cymbopogon citratus*, F. Poaceae), and Icacina plant (*Icacinatricantha*, F. Icacinaceae) are essential. Sandpaper plant (*Ficus exasperata*, F. Moraceae, *Ewe eepin* in Yoruba), Moringa (*Moringa oleifera*, F. Moringaceae), Levant cotton (Gossypium herbaceum, F. Malvaceae) and Achiote (*Bixa Orellana* Linn, family Bixaceae) are very vital. Furthermore, the vegetable section would also require a deliberate planning in the landscape framework of homegardens. Scent leaf plant (*Ocimumgratissimum,* F. Lsabiatae), Bitter leaf plant (*Vernonia amygdalina,* F. Asteraceae) and Fluted pumpkin (*Ugwu*) leaf (*Telfairia occidentalis,* F. Cucurbits) are recommended for home landscape design specification. Pepper (*Capsicum annuum*, F. Solanaceae), Water leaf (*Talinum fruticosum*, F. Talinaceae), Round pepper (*Capsiumfructescens*, F. Solanaceae), Ivy gourd (*Coccinia grandis*, F. Cucurbitaceae) are not to be left out.

Eco-cultural homegardens would benefit maximally from orchard landscape planning. A good mix and collections of Pineapple (*Ananas comosus*, F. Bromeliaceae), Passion fruit (*Passiflora edulis*, F. Passifloraceae), Cashew (*Anacardium occidentale*, F. Anacardiaceae), Sweet orange (*Citrus sinensis*, F. Rutaceae), and Plantain (*Musa paradisiaca*, F. Musaceae) are essential for inclusion in home landscape plant specification for building plan approval processing. They may also be specified for general landscaping around the home and not necessarily in a garden. Jackfruit (*Artocarpus heterophyllus*, F. Moraceae), lime (*Citrus aurantifolia*), lemon (*Citrus limon*, F. Rutaceae), mango (*Mangifera indica,* F. Anacardiaceae) and pawpaw (*Carica papaya,* F. Caricaceae) have multiple ecosystem services that make them indispensable for formal specification for general home landscape planning and homegarden design. Soursop (*Annona muricata,* F. Annonaceae), Guava (*Psidium guajava*, F. Myrtaceae), and Tangerine (*Citrus reticulata*, F. Rutaceae) could be specified as a triad of homegarden plants in the landscape design or general home landscapes.

Another approach to sustainable ecological urbanism in Yoruba cities is urban agriculture. Urban agriculture space is a viable eco-cultural component for formal crop garden planning in contiguity with home building. This is in view of its numerous ecosystem services, particularly for provisioning and micro-climate regulating ecosystem services that enhance indoor-outdoor thermal regulation and serving as wind breakers. In view of the unfavourable economic realities, urban agriculture in the home environment in Yoruba cities provides adequate economic leverage for the average family in small and medium size cities as evidenced in this study. Therefore, Cassava (*Manihot esculenta*, F. Euphorbiaceae), Cocoyam (*Xanthosoma saggitifolium*, F. Araceae), Yam (Dioscorea spp. – water, white, bitter, and yellow yam), and Maize (*Zea mays*, F. Poaceae) could be specified for formal home landscape and homegarden planning. If these approaches of producing urban green infrastructure are mainstreamed into the city planning activities such as urban design, urban renewal, and building and planning regulations, Yoruba cities would adequately blend IKS and modernism into unique eco-cultural milieu and opportunity for ecological branding of ecosystem services.

References

Adedeji JA, Amole SA (2010) Book review: Amos Rapoport, 1969. House form and culture (Foundations of Cultural Geography Series. In: Space and culture: the international journal of social spaces Englewood Cliffs, Prentice Hall, p 150. Retrieved from http://www.spaceandculture.orgrg/2010/07/15/book-review-house-form-and-culture/

Adedeji JA, Fadamiro JA (2018) Urbanisation forces on the landscapes and the changing value-systems of Osun Sacred Grove UNESCO Site, Osogbo, Nigeria. Landsc Res 43(6):798–816

Adedeji JA, Fadamiro JA, Odeyale TO (2020) Design toolkits for campus open spaces from post-occupancy evaluations of federal universities in South-west Nigeria. Built Environ Proj Asset Manag 10(2):296–311

Baschak LA, Brown RD (1995) An ecological framework for the planning, design and management of urban river greenways. Landsc Urban Plan 33(1–3):211–225

Burgess EW (1925) The growth of the city: An introduction to a research project. In: Park RE, Burgess EW, McKenzie R (eds) The city. University of Chicago Press, Chicago, pp 47–62

Claghorn J (2018) Algorithmic landscapes: computational methods for the mediation of form, information, and performance in landscape architecture. Gottfried Wilhelm Leibniz Universität, Hannover, Diss., 2018, 445 S. https://doi.org/10.15488/3734

Deleuze G, Guattari F (1994) What is philosophy? Columbia University Press, New York

Douglas I (2019) 50 years change in urban land use and ecological planning globally in the era of design with nature. Ecosyst Health Sustain 5(1):185–198

Elgin CZ (2017) Trueenough. MIT Press, Cambridge, MA\London

Gobo G (2018) Upside down – reinventing research design 1. In: The SAGE handbook of qualitative data collection. SAGE Publications Ltd, London, p 65–83

Grimm NB, Grove JG, Pickett ST, Redman CL (2000) Integrated approaches to long-term studies of urban ecological systems: urban ecological systems present multiple challenges to ecologists—Pervasive human impact and extreme heterogeneity of cities, and the need to integrate social and ecological approaches, concepts, and theory. Bioscience 50(7):571–584

Haase D, Jänicke C, Wellmann T (2019) Front and back yard green analysis with subpixel vegetation fractions from earth observation data in a city. Landsc Urban Plan 182:44–54

Hersperger AM, Grădinaru SR, Pierri Daunt AB et al (2021) Landscape ecological concepts in planning: review of recent developments. Landsc Ecol 36:2329–2345

Jabareen Y (2009) Building a conceptual framework: philosophy, definitions, and procedure. Int J Qual Methods 8(4):49–62

Johnson AH, Berger J, McHarg IL (1979) A case study in ecological planning: The Woodlands, Texas. Plan Uses Manage Land 21:935–955

La Rosa D, Takatori C, Shimizu H, Privitera R (2018) A planning framework to evaluate demands and preferences by different social groups for accessibility to urban greenspaces. Sustain Cities Soc 36:346–362

Lefebvre H, Nicholson-Smith D (1991) The production of space, vol 142. Blackwell, Oxford

Li FL, Xu YG (2021) Study on plant landscape planning method based on discrete particle swarm optimisation. Int J Environ Technol Manag 24(3–4):184–199

McHarg IL (1969) Design with nature. Natural History Press, New York. (2nd ed., John Wiley & Sons, New York, 1994)
Perrotti D, Stremke S (2020) Can urban metabolism models advance green infrastructure planning? Insights from ecosystem services research. Environ Plan B: Urban Anal City Sci 47(4):678–694
Prokopenko L (2020) Changing space. In: The Routledge companion to art in the public realm, Routledge, pp 257–268
Qtaishat Y, Adeyeye K, Emmitt S (2020) Eco-Cultural Design Assessment Framework and Tool for Sustainable Housing Schemes. Urban Sci 4(4):65
Rapoport A (1969) House form and culture. (Foundations of cultural geography series). Prentice Hall, Englewood Cliffs, pp 150
Seaver N (2017) Algorithms as culture: some tactics for the ethnography of algorithmic systems. Big Data Soc 4(2):2053951717738104
Shafizadeh-Moghadam H, Minaei M, Pontius RG Jr, Asghari A, Dadashpoor H (2021) Integrating a forward feature selection algorithm, random forest, and cellular automata to extrapolate urban growth in the Tehran-Karaj Region of Iran. Comput Environ Urban Syst 87:101595
Steiner F (2008) An ecological approach to landscape planning. Island Press, Washington DC
Tandarić N, Ives CD, Watkins C (2020) Can we plan for urban cultural ecosystem services? J Urban Ecol *6*(1):juaa 016
Tansley AG (1935) The use and abuse of vegetational concepts and terms. Ecology 16:284–307
Wang Q, Zeng X (2020) Energy-saving optimisation method for green space planning of urban gardens based on artificial bee colony algorithm. Int J Glob Energy Issues 42(5–6):393–408
Yang B, Li S (2016) Design with Nature: Ian McHarg's ecological wisdom as actionable and practical knowledge. Landsc Urban Plan 155:21–32. https://doi.org/10.1016/j.landurbplan.2016.04
Yılmaz S, Özgüner H, Mumcu S (2018) An aesthetic approach to planting design in urban parks and greenspaces. Landsc Res 43(7):965–983

Index

J. A. Adedeji, *Ecological Urbanism of Yoruba Cities in Nigeria*, Cities and Nature, https://doi.org/10.1007/978-3-031-34688-0

T